Johns Hopkins Atlas of Human Functional Anatomy

Original Illustrations, with Descriptive Legends, by

Leon Schlossberg

Text Edited by **George D. Zuidema, M.D.**

Fourth Edition

The Johns Hopkins Atlas of Human Functional Anatomy

The Johns Hopkins University Press ▪ Baltimore and London

To Jean and Joan

Plates 2, 25, 35, and 46, courtesy of Winthrop Laboratories; plates 3, 4, and 5, courtesy of the Medical Models Laboratory; plates 11, 12, 13, 18, 30, 57, 58, 59, 60, 61, 62, 63, 64, 65, 66, 67, 68, 71, 72, 73, 74, 75, 76, and 77, all originally published by W. B. Saunders Company; plates 14 and 15, courtesy of the Medical Department, U.S. Navy. Plate 22 first appeared in *Atlas of Aortic Surgery,* by G. Melville Williams and Leon Schlossberg (Baltimore: Williams & Wilkins, 1997). Plate 41 first appeared in *Atlas of General Thoracic Surgery,* by Mark M. Ravitch and Felician M. Steichen (Philadelphia: W. B. Saunders Company, 1988). Plates 49, 50, 51, and 52 first appeared in *The Operative Management of Breast Disease,* by R. Robinson Baker and John E. Neiderhuber (Philadelphia: W. B. Saunders Company, 1992). Plate 56 first appeared in *Atlas of Nuclear Medicine,* vol. 2, by Frank H. Deland and Henry N. Wagner (Philadelphia: W. B. Saunders Company, 1970). All are reproduced with permission.

 First edition 1977
Fourth edition 1997

Printed in the United States of America on acid-free paper
06 05 04 03 02 01 00 99 98 97 5 4 3 2 1

The Johns Hopkins University Press
2715 North Charles Street
Baltimore, Maryland 21218-4319
The Johns Hopkins Press Ltd., London

Library of Congress Cataloging-in-Publication Data will be found at the end of this book.

A catalog record for this book is available from the British Library.

ISBN 0-8018-5651-5
ISBN 0-8018-5652-3 (pbk.)

Contributors

The Johns Hopkins University School of Medicine

FACULTY

R. Robinson Baker, M.D. Professor of Surgery and Professor of Oncology

William R. Bell, M.D. Professor of Medicine, Professor of Radiology, and Professor of Nuclear Medicine

Robert K. Brawley, M.D. Associate Professor of Surgery

Gregory B. Bulkley, M.D. Mark M. Ravitch Professor of Surgery

Michael A. Choti, M.D. Assistant Professor of Surgery

Rainer M. E. Engel, M.D. Associate Professor of Urology

Margaret M. Fletcher, M.D. Assistant Professor of Otolaryngology/Head and Neck Surgery

Vincent L. Gott, M.D. Richard Bennett Darnall Professor of Surgery

David W. Heese, D.D.S. Assistant Professor of Dental Surgery

Thomas R. Hendrix, M.D. Professor of Medicine

Charles E. Iliff, M.D. Professor Emeritus of Ophthalmology

Donlin M. Long, M.D., Ph.D. Professor of Neurological Surgery and Chair of the Department of Neurological Surgery

George T. Nager, M.D. Andelot Professor Emeritus of Otolaryngology/Head and Neck Surgery

James J. Ryan, M.D. Assistant Professor of Plastic Surgery

Leon Schlossberg Associate Professor of Art as Applied to Medicine

Mark A. Talamini, M.D. Assistant Professor of Surgery

George B. Udvarhelyi, M.D. Professor Emeritus of Neurological Surgery and Associate Professor Emeritus of Radiology

Henry N. Wagner, Jr., M.D. Professor and Director of the Division of Nuclear Medicine and Radiation Health

Patrick C. Walsh, M.D. Professor and Director of the Department of Urology

FORMER FACULTY

Melvin H. Epstein, M.D. Professor and Co-Chair of the Department of Clinical Neurosciences, Brown University, Providence, Rhode Island

Donald S. Gann, M.D. Professor and Executive Vice Chair of the Department of Surgery, University of Maryland School of Medicine, Baltimore, Maryland

James L. Hughes, M.D. Professor of Orthopedic Surgery and Chair of the Department of Orthopedic Surgery, University of Mississippi Medical Center, Jackson, Mississippi

James P. Isaacs, M.D. Salisbury, Maryland

Howard W. Jones, Jr., M.D. Professor of Obstetrics and Gynecology, Eastern Virginia Medical School, Norfolk, Virginia, and Professor Emeritus, Johns Hopkins University School of Medicine

H. Lorrin Lau, M.D. Former Chair of the Department of Obstetrics and Gynecology, St. Francis Medical Center West and Kuakini Medical Center, Honolulu, Hawaii

Vernon T. Tolo, M.D. Professor of Orthopedic Surgery, University of Southern California, Los Angeles, California

John J. White, M.D. Clinical Professor, Department of Surgery and Department of Pediatrics, Mercer University School of Medicine, Macon, Georgia

George D. Zuidema, M.D. Professor of Surgery and Former Vice Provost for Medical Affairs, University of Michigan, Ann Arbor, Michigan

Contents

PLATES

Preface to the Fourth Edition

This volume unites illustrations created by the unique artistic talents of Leon Schlossberg with a text written by the faculty of the Johns Hopkins University School of Medicine. The book itself is designed to present basic anatomy to students of medicine and the allied health professions as well to undergraduates; it does so in a manner that emphasizes function as well as structure. The text covers basic principles, with sufficient attention to detail to permit students to orient themselves to the subject and obtain a working knowledge of anatomical systems and specialized organs. The functional approach helps the reader to understand interrelationships.

With each new edition we have introduced new material and have refined the accompanying text. For the fourth edition we added sixteen new plates and text illustrating points of important clinical interest. For example, there is new coverage of the breast and its lymphatic drainage areas; a new description of the segmental anatomy of the liver; new material on the prostate gland and its nerve supply; and a new section on abdominal anatomy as viewed for laparoscopic surgery. In addition, we have added a large anatomical display of a man which emphasizes areas of clinical interest. The result is an improved, more extensive volume, which should serve as a useful handbook and a departure point for the detailed study of anatomy by other techniques.

In certain instances, newer anatomical nomenclature differs from that generally used by clinicians. Where this situation occurs, we have elected to use the terms that are used most widely by physicians and surgeons in clinical practice, for it is this terminology the student is also most likely to use. The artist and the authors believe that this book will be a useful introduction to the initial phase of health science education.

George D. Zuidema, M.D.

Acknowledgments

I am indebted to Dr. George D. Zuidema for his interest in the idea for this book and for his help in bringing it to fruition. To the physicians and surgeons in many of the specialties at the Johns Hopkins University School of Medicine go most sincere thanks for the text they contributed and for the untiring manner in which they cooperated in consultations and in offering their criticisms. It is a distinct privilege for a medical illustrator to have access to the authoritative advice of anatomical, medical, and surgical experts. I owe a personal debt to the late Max Broedel, founder of the Department of Art as Applied to Medicine, for it was he who almost single-handedly created at the Johns Hopkins Medical Institutions a niche for medical illustration, from which has developed an excellent professional relationship between physician and illustrator.

I would like to thank the Medical Department of Winthrop Laboratories for their generous permission to use several of the illustrations that were originally produced for them. The W. B. Saunders Company kindly allowed me to reuse a number of my plates from the Schlossberg-Zuidema *Atlas of Surgical Anatomy of the Abdomen and Pelvis* and one of my illustrations from Deland and Wagner's *Atlas of Nuclear Medicine*. To the Medical Department of the U.S. Navy I extend grateful acknowledgment for the use of the plates on blood I created for them, and to the Medical Models Laboratory I offer thanks for allowing me to reproduce here the comprehensive portrayal of the skeletal anatomy that they originally commissioned from me.

To my wife, Jean, who encouraged me from the outset of this project, I am especially grateful. She skillfully translated my notes into understandable labels, with which she personally prepared the initial overlays for the illustrations.

To Wendy A. Harris, medical editor at the Johns Hopkins University Press, goes my most sincere gratitude for her understanding of the details involved in this major revision of the *Atlas*. To Anita Walker Scott, design and production manager at the Johns Hopkins University Press, I likewise wish to acknowledge my gratitude for her expertise in the preparation of the labels and illustrations for publication.

Leon Schlossberg

Introduction

D*orland's Medical Dictionary* defines anatomy as: "1. The science of the structure of the animal body and the relation of its parts. It is largely based on dissection, from which it obtains its name. 2. Dissection of an organized body." The definition continues to include applied anatomy, as well as artificial, artistic, microscopic, classic, comparative, general, gross, biologic, medical, pathologic, physiologic, and other subdivisions of the study of anatomy. Physiology is defined in Dorland's as: "The science which treats of the functions of the living organism and its parts." It is this area—functional living anatomy and the component structures of the body and the relation of its parts—that this work will endeavor to depict.

The key illustration in the chapter on the central nervous system is entitled "The Five Senses of Consciousness": hearing, seeing, tasting, smelling, and sense of movement and position of the body. These faculties are defined in accompanying legends and are further elucidated in the text describing the anatomy and functions of the brain, nerves, and spinal cord. Other body functions treated in like manner include skeletal, articulatory, muscular, alimentary, respiratory, urinary, reproductive, lymphatic, endocrine, integumentary, and autonomic and somatic nerve functions, and blood and cerebrospinal fluid circulation. The activity of these systems represents total body function. The organs—their structure and their topographical relationships—are illustrated and defined by the captions and accompanying descriptive text. Concerted efforts have been made to present the anatomy and function to facilitate the understanding of clinical conditions and diagnostic and corrective procedures. Since living anatomy and function are depicted, some of the study and reference material utilized includes radiopaque studies of living persons and surgical demonstrations of topographical relationships.

Scientific terms refer to position, portions, regions, and spaces of the body, including the head, neck, thorax, abdomen, pelvis, and upper and lower extremities. External anatomical positions include the following aspects: front, anterior (ventral); back, posterior (dorsal), and sides, lateral. Cross sections shown are midsagittal (along a line in the midline from front to back), sagittal (along any line parallel to midsagittal), horizontal or transverse (on a plane perpendicular to the sagittal plane), and frontal or coronal (on a plane from side to side). A view from above the head or toward the head is termed *cephalic* or *cephalad.* A view from below the coccyx or toward the coccyx is termed *caudal* or *caudad. Superior* is synonymous with *cephalic,* and *inferior* is synonymous with *caudal.*

Leon Schlossberg

Plate 1.
Anatomic Positions, Surface and Topographic Anatomy, and Regions and Planes of Sections

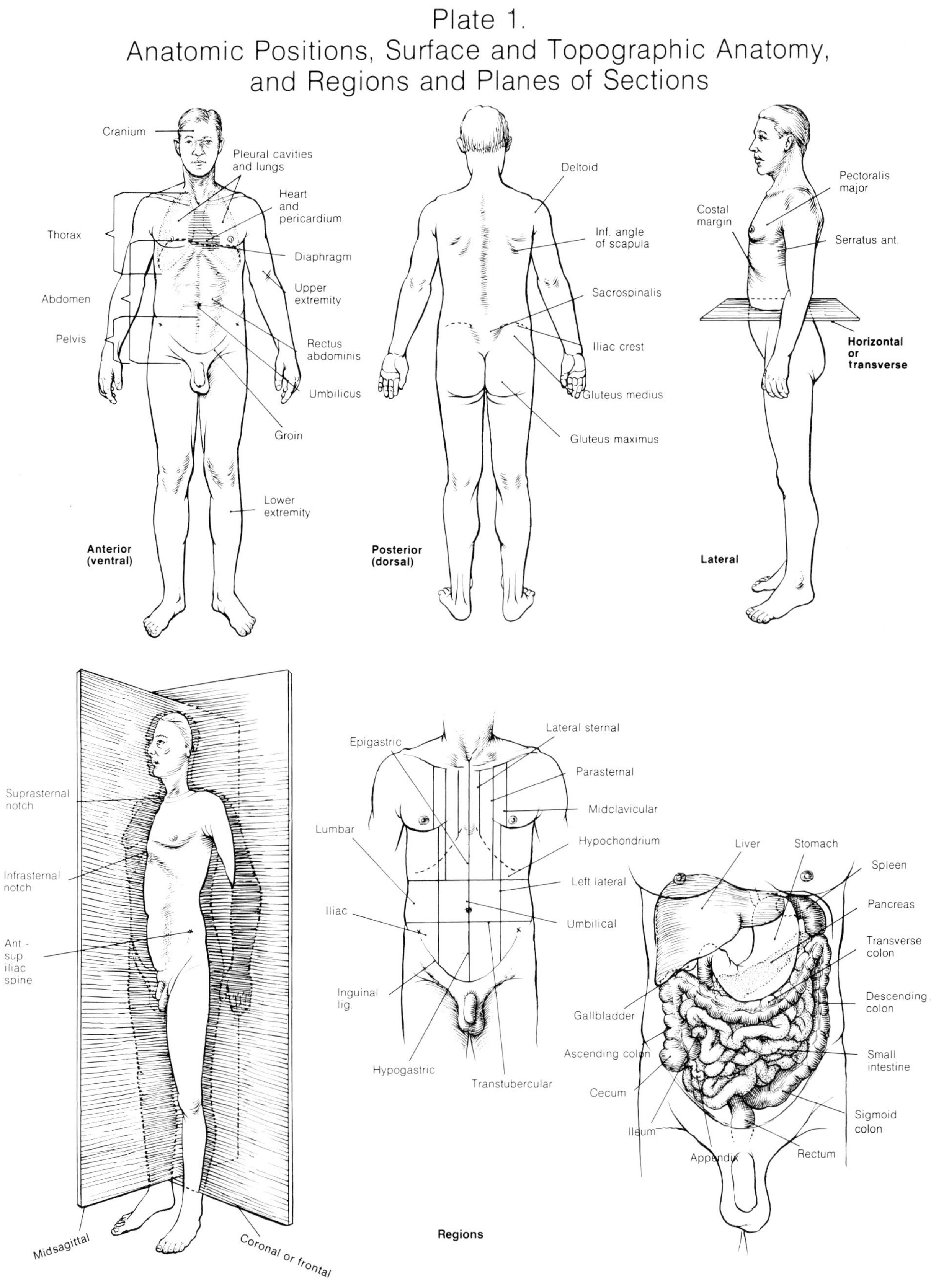

The Johns Hopkins Atlas of Human Functional Anatomy

Fetal Circulation

1

John J. White, M.D.

The basic requirements for human life are nutrition and respiration. These requirements are even more special in the fetus, which, starting from the union of two germ cells and emerging forty weeks later as a neonate, must undergo formation, development, and growth. The requirements in question are provided in utero via the fetal circulatory system.

During its entire intrauterine existence, the fetus is, in effect, a parasite dependent completely upon its mother for its nutrition and respiration. Shortly after embryogenesis begins, the essence of the fetal circulatory system is formed by the placental-umbilical circulation. The placenta, an organ that is made up of both fetal and maternal uterine contributory parts, adheres firmly to the inner wall of the uterus. On the maternal side, the uterine circulation hypertrophies significantly, and specialized end vessels elaborate in the decidua basalis, the maternal contribution to the placenta. On the fetal side of the placenta, a similar network of end vessels is elaborated in the chorionic villi.

The two circulations, maternal uterine and fetal, remain separate. The spaces between them are called *intervillous spaces* (see Plate 2, detail). Nutrients provided by the maternal circulation diffuse across the uterine capillary membrane, to be picked up, across the membranes of the umbilical capillary network, by the fetal circulation. Similarly, oxygen is taken up by the fetus and carbon dioxide is released.

From the placenta, the oxygen-rich, nutritious blood is delivered to the fetus by the umbilical vein in the umbilical cord. Upon entry to the fetus at the umbilicus, the umbilical vein travels to the liver, where it divides into two branches. A lesser portion of the blood supplies the liver, via the sinus intermedius. Most of the oxygenated and nutrient-rich blood passes through the liver, via the ductus venosus, to the inferior vena cava and the right heart. In the right atrium it mixes somewhat with cephalic venous blood of the fetus and is then shunted past the nonfunctional fetal lungs by two pathways basically. In the heart, most of the oxygenated blood flows through an opening between the two atria, the foramen ovale, into the left heart for distribution to the entire body (red arrows). Most of the unoxygenated fetal venous blood flows from the right atrium into the right ventricle and pulmonary artery (blue arrows). At this point another shunt, the ductus arteriosus, allows this blood to flow into the descending aorta, bypassing the brain and heart. Deoxygenated and nutrient-depleted blood from the fetus flows via major branch vessels of the aorta, the paired internal iliac or hypogastric arteries, back to the umbilicus, where these arteries pass, in the cord as the umbilical arteries, back to the placenta.

With and immediately after birth, physiologic processes active in the mother and baby bring about changes from the fetal-placental circulation to the normal extrauterine circulatory state. Respiratory activity of the neonate clears fluid from the lungs, aerates and inflates the alveoli, and stimulates active blood flow in the pulmonary vessels. As the pulmonary vascular resistance lowers, the decreased pressure at the ductus arteriosus decreases its size; functionally, it generally closes shortly after birth. As the pressure in the left atrium builds up with blood return from the lungs, the foramen ovale also closes functionally. This is abetted by diminished pressure in the right atrium as the ductus venosus also ceases to shunt blood through the liver. The hypogastric arteries obliterate back to the internal iliacs, which then only supply the pelvis.

Plate 2.
Fetal Circulation

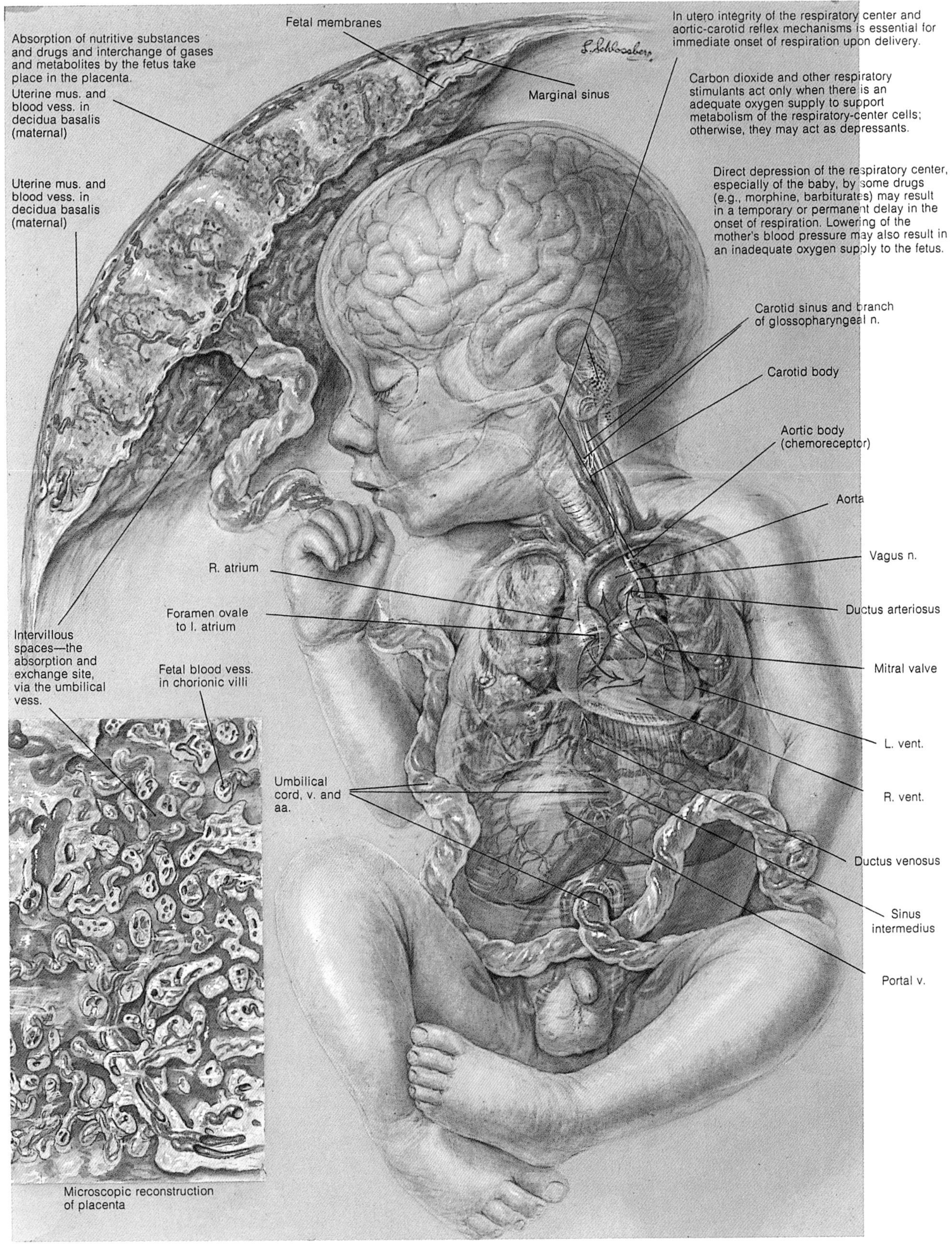

Skeletal Anatomy

2

James L. Hughes, M.D.

The skeletal system is formed from bone, a complex, specialized connective tissue. Bone has many unique properties that enable it to carry out a diversity of functions. It performs a mechanical function in providing for the skeletal support of the body, in protecting the vital organs of the cranial and thoracic cavities, and in providing the platform for the attachment of muscles. The structure of bone is beautifully adapted to its various roles by utilizing the least material in association with the least weight that will enable it to function properly. Bone is engineered like reinforced concrete, with the collagen fibers resembling steel rods and the calcium the concrete itself.

Bone is active metabolically and provides for the immediate calcium needs of the body. The blood calcium concentration is maintained at a steady level by the reciprocal activity of parathyroid hormone and thyrocalcitonin on the bone cells. These cells regulate bone matrix and bone mineral (hydroxyapatite crystals) formation and resorption. Bone mineral can be added or removed from the matrix, depending on the body's need for calcium. In the normal adult, the formed elements of the blood—red cells, white cells, and platelets—originate in the bone marrow cavity. The principal source of these hematopoietic cells is the long bones (red marrow), while the marrow cavities (yellow marrow) of many other bones exist as reserve sites in the time of need.

Macroscopically, bone is either cancellous (spongy) or compact. Cancellous bone is primarily formed in the flat bones of the body—that is, in the cranium, pelvis, vertebrae, and scapula—and is rapidly responsive to metabolic exchanges. The cancellous bone occupies significant space in the skeleton, but it accounts for only one-fifth of the mass of bone. Both cancellous and compact bone consist of the same materials but differ in their architecture. Cancellous bone is arranged in trabeculae (thin plates), while compact bone is arranged in a lamellar pattern formed by concentric layers of bone fibers. This arrangement of fibers in lamellar bone provides maximum strength; thus, lamellar bone is found primarily in the long bones, which are constantly under stress.

Lamellar bone consists of numerous Haversian systems. A Haversian system, or osteon, consists of a central blood vessel, bone cells, and lamellar bone.

Microscopically, the greatest part of the mass of bone is made up of bone matrix. Embedded within this matrix are lacunae (cavities), which are completely filled with bone cells called *osteocytes*. Originating from the lacunae are small canals called *canaliculi*, which penetrate the hard interstitial substance in all directions. These canals branch and communicate with each other in such a manner as to connect all the lacunae together. The presence of this network of canals, plus the fact that no bone cell is more than 0.1 mm from a capillary, guarantees that every cell in bone is nourished and viable.

Remodeling and rebuilding of bone is accomplished by cells found on the surface of bone, the osteoclasts and osteoblasts. All three types of bone cells arise from a common stem cell, and transformation from one to the other is frequent. The activities of these cells enable a fractured bone to heal and remodel in accordance with the stresses applied to it.

In a typical long bone, such as the femur, the diaphysis (shaft) consists of compact bone surrounding a large cavity (medullary cavity) housing the bone marrow. The ends of the long bones are called *epiphyses* and consist of cancellous bone surrounded by a thin layer of cortical bone. In a growing child, the diaphysis is separated from the epiphysis by the epiphyseal cartilage plate. This area of trabecular bone and cartilage plate is sometimes called the *metaphysis* and permits

Plate 5.
Skeleton—Lateral View

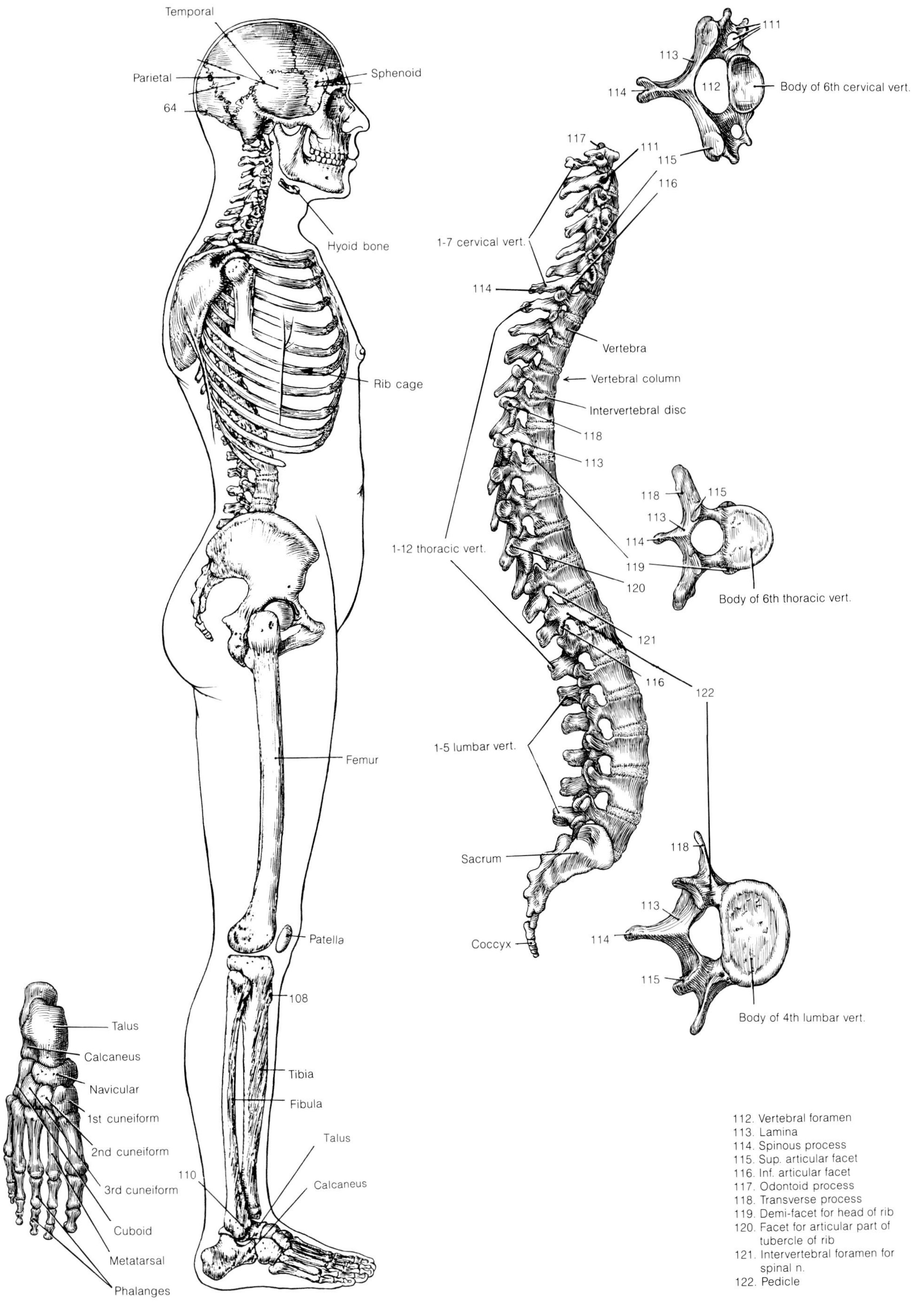

Temporal
Parietal
64
Sphenoid
Hyoid bone
Rib cage
Femur
Patella
108
Tibia
Fibula
Talus
Calcaneus
110
Talus
Calcaneus
Navicular
1st cuneiform
2nd cuneiform
3rd cuneiform
Cuboid
Metatarsal
Phalanges
111
113
112
114
Body of 6th cervical vert.
115
117
111
116
1-7 cervical vert.
114
Vertebra
Vertebral column
Intervertebral disc
118
113
118
115
113
114
1-12 thoracic vert.
119
120
Body of 6th thoracic vert.
121
116
122
1-5 lumbar vert.
Sacrum
Coccyx
118
113
114
115
Body of 4th lumbar vert.

112. Vertebral foramen
113. Lamina
114. Spinous process
115. Sup. articular facet
116. Inf. articular facet
117. Odontoid process
118. Transverse process
119. Demi-facet for head of rib
120. Facet for articular part of tubercle of rib
121. Intervertebral foramen for spinal n.
122. Pedicle

Skeletal Muscles, Joints, and Fascial Structures

Vernon T. Tolo, M.D.

The essential feature of muscle, which leads to movement in all parts of the body, is its ability to contract. Unlike smooth muscle or cardiac muscle, skeletal, or striated, muscle contracts only under nerve control, i.e., voluntarily or reflexly. This muscle type accounts for about 40 percent of human body weight. Although Leonardo da Vinci in his classic anatomical drawings used letters to designate specific muscles, and Galen and Vesalius used numbers for this purpose, muscles are currently referred to by names. These names may describe the function, direction, or attachment of the muscles, or may simply describe the shape, size, or structure of the specific muscle.

Muscles are composed of multiple bundles of fibers. Hypertrophy of existing muscle fibers occurs with muscle training, but there is no way to increase the number of muscle fibers. Muscle must be stressed to work effectively, and inactivity or bedrest leads to a loss of muscle size and strength.

The muscle attachments are usually designated the origin and the insertion. The origin is located proximally and near the midline, while the insertion is the distal, peripheral attachment. These attachments bridge one or more joints, and the result of muscle contraction is joint movement. The principal actions of muscles are fourfold. A prime mover muscle effects the actual movement that occurs. An antagonist acts by relaxing to enable the prime mover to work, and the antagonist is capable, by contraction, of preventing the movement. A fixation muscle provides a stable base by steadying a part of the body from which other muscles perform, and a synergistic muscle controls the position of intermediate joints so that the prime movers passing over several joints may exert power in moving the distal joint.

Each group of several fibers receives a nerve supply that allows voluntary contraction of the muscle. Initiated in the cerebral cortex, the stimulus for voluntary muscle action passes down the spinal cord and the nerve root to the motor end-plate, the junction of the nerve with the muscle fibers. Release of acetylcholine at the motor end-plate allows transmission of the stimulus from the nerve to the muscle, resulting in the contraction of the muscle fibers. When a muscle contracts, all fibers innervated by the stimulated motor end-plate contract either fully or not at all, with increasing numbers of fibers recruited as the work load becomes heavier. During muscle action, the basic microscopic muscle unit, the sarcomere, shortens. The energy for the contraction of the muscle is obtained primarily from free fatty acids, glucose, or glycogen. Muscles requiring rapid movement largely contain white, or fast, fibers high in glycogen content, and muscles doing sustained, heavier work are of the red, or slow, fiber type. These fiber groups are found in varying amounts in different muscles depending on the primary muscle function.

Prolongations of the muscles at the insertion are called *tendons,* collagenous tissue bands that obviate the need for long muscle fibers, decrease the bulk of tissue around the joints, and provide protection against muscle injury during sudden movement of the joint. Many tendons are enclosed in a synovial sheath, a fascial structure that provides smoother motion and nutrition for the tendons.

Fascial structures are present throughout the body and are derived from the mesoderm, the same embryonic tissue from which muscles arise. Superficial fascia, lying just beneath the skin, is impregnated with fat of differing degrees. Because of its areolar quality, this layer is particularly subject to edema following limb injury. The deep fascia is a thick fibrous membrane that encloses muscles, blends with the ligaments, and attaches to bony surfaces. By forming intermuscular septa, the deep fascia separates groups of muscles or individual muscles. By forming retinaculae at the wrist and ankle, it keeps the tendons in position and acts as a pulley for tendon action. The synovial sheaths are localized thickenings of the deep fascia to facilitate tendon movement. Bursae are sacs of synovial tissue that develop between a moving tendon and a bony prominence or ligament over which the tendon passes. They are usually located near joints. Bursitis, the inflammation of these structures, is a common ailment in the shoulder, elbow, and knee after excessive or unusual activity of these joints.

Joints are the articulations between adjacent bones. The majority of the joints of the extremities are synovial joints, which are specialized to allow free movement. The synovium lines the joint cavity and secretes synovial fluid to nourish smooth, avascular articular cartilage covering the ends of the articulating bones. The articulating bones remain opposed by muscle tension and the strength of ligaments bridging the joints. Ligaments are fibrous tissue bands that cross a joint and change lengths with joint movement. They do not voluntarily contract or relax. Ligaments attach bone to bone, whereas tendons attach muscle to bones. Two common joint types are the hinge joint (elbow, knee, and finger) and the ball-and-socket joint (hip and shoulder).

NECK AND TRUNK MUSCLES

The axial muscles of the neck and trunk are innervated by spinal nerves and are segmentally arranged. The largest muscles involved with neck movements are the sternomastoid and the trapezius. Additional anterior neck muscles are the longus colli, scalenus anterior, scalenus medius, scalenus posterior, rectus capitis, and longus capitis. The platysma muscle, with a role in facial expression, lies anteriorly in the region of the superficial fascia. Swallowing, speech, and chewing involve the muscles in the laryngeal and esophageal regions called *strap* muscles, which include the omohyoid, sternohyoid, sternothyroid, thyrohyoid, digastric, stylohyoid, mylohyoid, and geniohyoid muscles.

The most important muscle in the thorax is the diaphragm, the principal muscle for respiration. Assisting the diaphragm in respiration are the external and internal intercostal muscles. Other thoracic muscles are the transversus thoracis, subcostals, levatores costarum, and serrati posteriores. The abdominal wall muscles, discussed in Chapter 4, occur in several layers, with the principal ones being the external oblique, internal oblique, transversus abdominus, and rectus abdominus muscle. The muscles of the back are arranged in three layers: the erector spinae system, the transversospinalis system, and the deep layer. The erector spinae group, consisting of the spinalis thoracis, longissimus, splenius capitis, and erector spinae muscles, runs longitudinally and spans several vertebrae. The transversospinalis system runs obliquely and consists of the multifidus and semispinalis muscle groups. These usually span fewer vertebrae. The deep layer includes the interspinalis, intertransverse, and suboccipital muscles, which span only adjacent vertebrae. The disabling effect of low back pain visibly demonstrates the interplay necessary between these multiple muscles if smooth functioning of the back is to occur. Asymmetrical function of the back muscles, particularly in neuromuscular diseases, may lead to scoliosis, a lateral curvature of the spine.

The most distal trunk muscles are those of the pelvic diaphragm and perineum. Necessary because of the erect posture in humans, the pelvic diaphragm holds up the pelvic floor and consists of the coccygeus and levator ani, formed by the iliococcygeus and pubococcygeus muscles. In the perineum, the anal triangle, located posteriorly, contains the sphincter ani externus; the urogenital triangle, situated anteriorly, contains the superficial transversus perinei, bulbospongiosus, ischiocavernosus, sphincter urethrae, and the deep transverse perineal muscles. The perineal body connects the anal and urogenital triangles.

Another upper trunk muscle group serves to connect the upper limb to the trunk. These muscles are responsible for scapular and clavicular motion as a part of upper extremity movement. The dorsal muscles are the trapezius, levator scapuli, rhomboids, and the latissimus dorsi, the widest muscle of the back. The ventral or anterior muscles are the pectoralis major, pectoralis minor, subclavius, and serratus anterior. Although the principal action of these muscles is involved with shoulder girdle motion, stabilization of the upper extremities will allow the reverse action to occur, and the pectoralis major and minor, latissimus dorsi, trapezius, and sternomastoid muscles can all work as accessory respiratory muscles.

UPPER EXTREMITY MUSCLES AND JOINTS

The shoulder, a ball-and-socket joint with the widest range of motion of any joint in the body, has freedom of motion at the expense of stability and is the most easily dislocated of the major joints. The shoulder girdle moves at the articulation between the humerus and glenoid cavity of the scapula, but the scapula also moves on the chest wall and the clavicle rotates at its joints. In abduction of the shoulder, for example, approximately two-thirds of this abduction results from movement between the glenoid and the humeral head, while one-third comes from rotation of the scapula on the chest wall. The glenoid is approximately one-third the surface of the humeral head. What strength and stability of the joint is present results from the attachments of muscles close to the articular surface. The least joint support occurs inferiorly, the direction in which dislocation most commonly occurs. Bursitis in

Plate 6. Muscles, Ligaments and Fasciae, Tendons, and Bursae—Anterior View

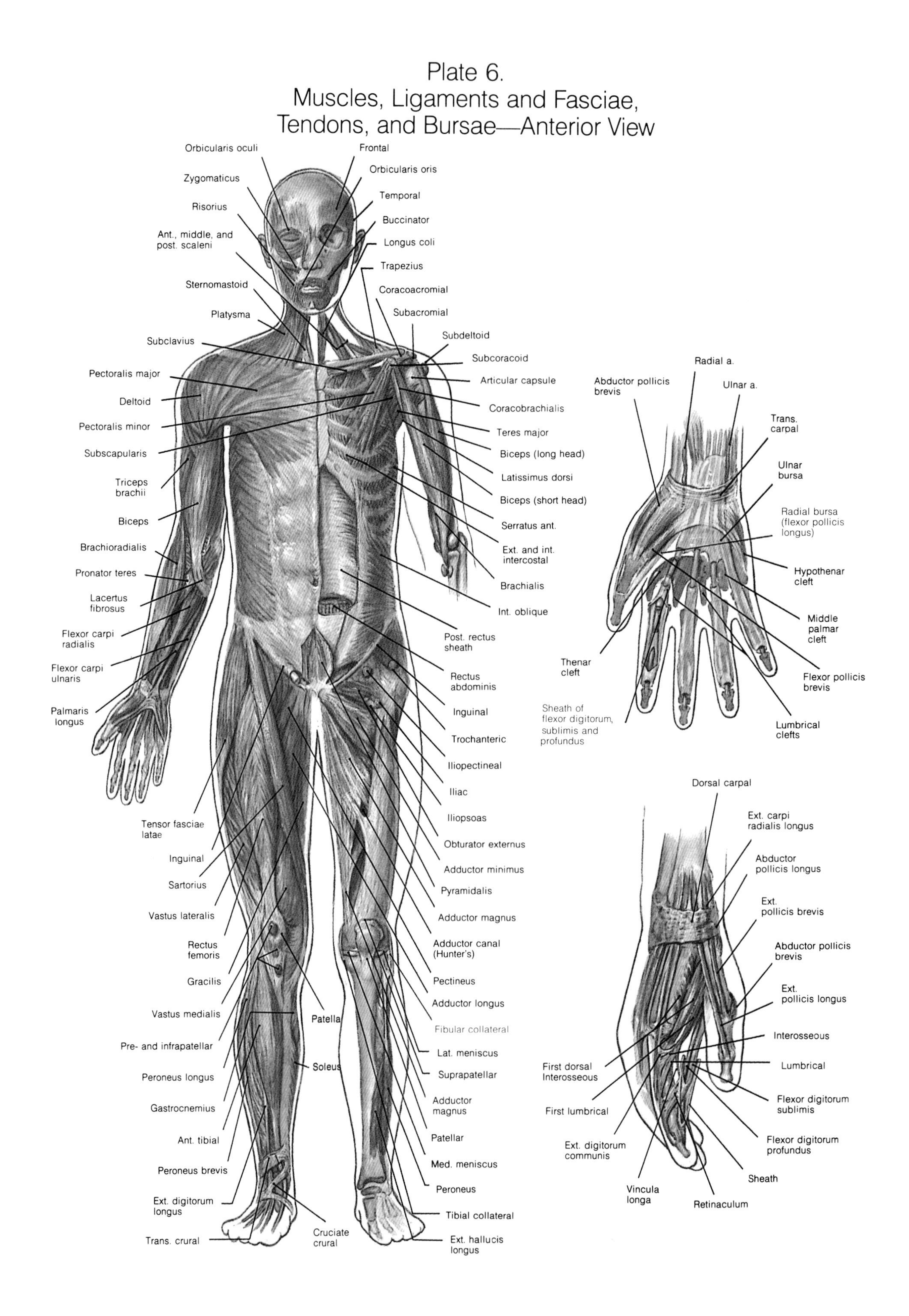

Plate 7. Muscles, Ligaments and Fasciae, Tendons, and Bursae—Posterior View

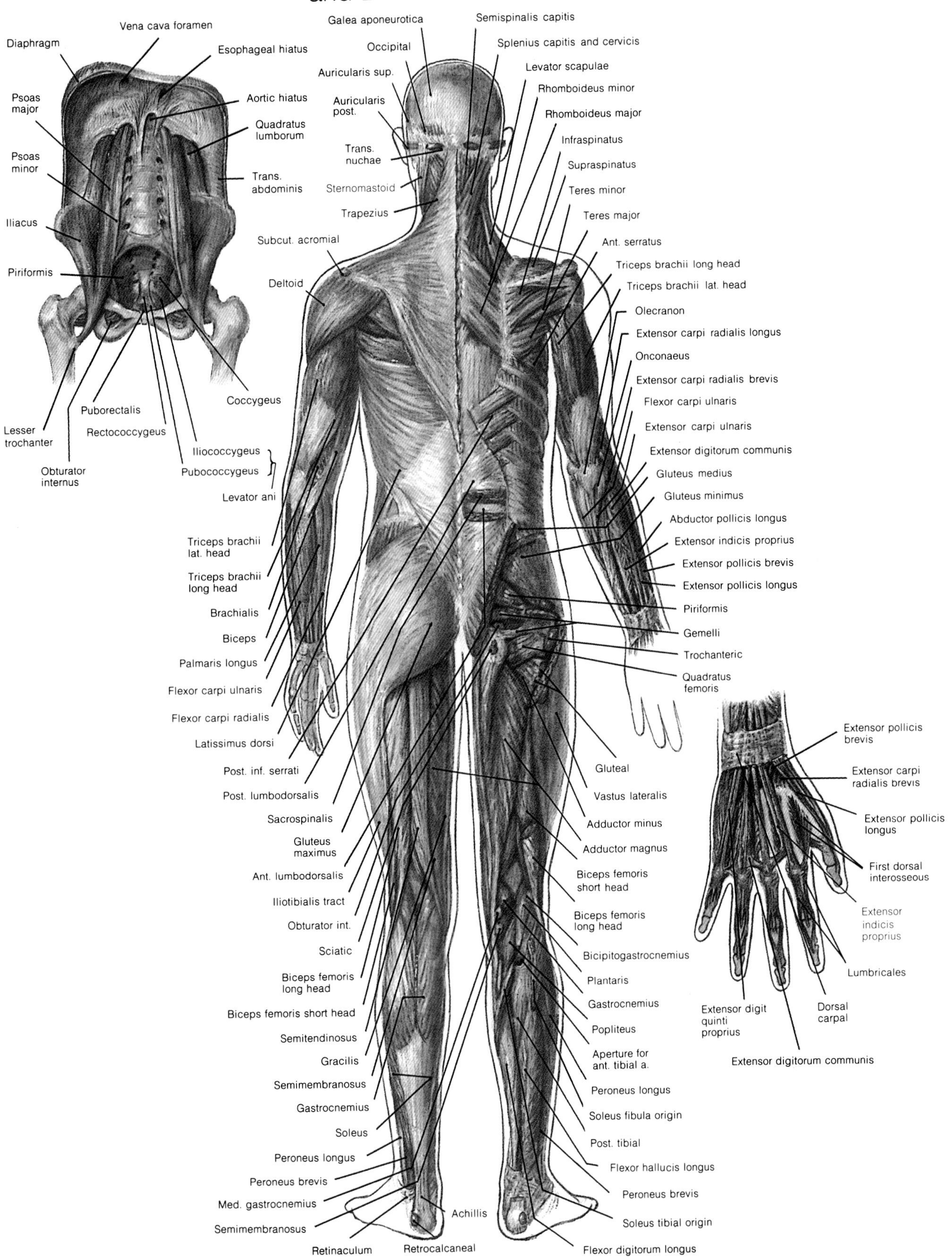

the shoulder area is common. The subscapular bursa communicates with the joint, whereas the subacromial bursa, located below the deltoid, does not.

Numerous muscles interact in shoulder motion, as the shoulder joint is free to move in almost any plane. The deltoid and rotator cuff are the principal abductors of the shoulder. The rotator cuff, consisting of subscapularis, supraspinatus, infraspinatus, and teres minor, keeps the articular surfaces together and initiates abduction, while the deltoid provides the major strength for continued abduction at the shoulder. Adduction and medial (internal) rotation of the shoulder are mediated by the pectoralis major, latissimus dorsi, and teres major. The infraspinatus and teres minor effect lateral rotation, while extension is performed by the latissimus dorsi and the posterior deltoid muscle. Flexion of the shoulder is mediated by the pectoralis major and the anterior deltoid fibers. The biceps and the triceps of the upper arm have separate origins, which cross over the shoulder joint but act weakly on shoulder motion per se.

The muscles of the upper arm are divided into anterior and posterior groups. The anterior group includes the coracobrachialis, brachialis, a pure flexor of the elbow, and biceps brachii, which functions primarily to supinate the forearm and secondarily to flex the elbow. The triceps brachii, for elbow extension, is posterior. The elbow is principally a hinge joint. The ulna articulates with the trochlea, while the radius articulates with the capitulum portion of the humerus. The collateral ligaments and muscles around the elbow augment the inherent stability of this joint provided by the bony contours of the humerus, ulna, and radius. The principal bursa in this area is that overlying the olecranon process of the ulna.

The muscles of the forearm form two groups. The anteromedial, flexor-pronator group is divided into three layers. The superficial layer has its origin on the medial humeral epicondyle and is formed by the pronator teres, flexor carpi radialis, palmaris longus, and flexor carpi ulnaris muscles. The second layer is formed by the flexor digitorum profundus, flexor pollicis longus, and, in the distal forearm, the pronator quadratus muscles. The posterolateral group forms two layers. The most superficial are the extensor carpi radialis longus, brachioradialis, anconeus, extensor carpi radialis brevis, extensor digitorum, extensor digiti minimi, and extensor carpi ulnaris. The last four muscles begin at the lateral humeral epicondyle; inflammation over this area constitutes "tennis elbow." The deep posterolateral layer is formed by the supinator, the abductor pollicis longus, the extensor pollicis brevis, extensor pollicis longus, and extensor indicis muscles. Movement at the wrist joint is principally in the plane of flexion and extension, but there is a complex interplay between the metacarpals, the eight carpal bones, and the distal radial and ulnar articulations.

The principal function of the upper extremity is to position the hand properly to do work or to guard or cover any part of the body. A major feature of human hand function is the ability to oppose the thumb to the digits. The interphalangeal joints of the hands are hinge joints, while the metacarpophalangeal joints allow some adduction and abduction as well as the hinge function. The first metacarpal joint is a saddle joint.

Movement of the hand is produced by a combination of the extrinsic muscles, originating in the forearm but inserting into the hand and wrist, and the shorter intrinsic muscles that arise within the hand itself. The thenar eminence consists of the abductor pollicis brevis, the opponens pollicis, and the flexor pollicis brevis muscles, all innervated by the median nerve. The adductor pollicis attaches deep to these muscles. The interosseous muscles include the palmar and dorsal groups. The four palmar interossei act as adductors from the hand at midline. The four dorsal interossei act as adductors from the midline with two muscular heads originating from adjacent metacarpal bones. The interossei also flex the metacarpophalangeal joints and extend the interphalangeal joints. The four lumbrical muscles, associated closely with the flexor digitorum profundus tendons, are coordinated with the interossei for fine finger movements. The ulnar nerve innervates all the intrinsic muscles of the hand except the lateral two lumbricals and the thenar eminence muscles.

On the dorsum of the hand the skin is mobile, and swelling occurs in the superficial fascia here more readily than in the palmar aspect, where the skin is firmly bound to the underlying fascial structures. The arrangement of the deep fascia in the hand forms potential spaces, especially the mid-palmar space deep to the flexor tendons, where infection may occur. Other sites of potential infection are the tendon sheath of the flexor pollicis longus, the individual flexor sheaths of the second, third, and fourth fingers, and the ulnar bursa, or common flexor tendon sheath, which communicates with all flexor tendons except the thumb. Closed-space infections in these areas can destroy the lining of the synovial sheath and limit normal tendon motion.

LOWER EXTREMITY MUSCLES AND JOINTS

The muscles and joints of the lower extremities provide strength and stability for the body. They serve to transmit the weight of the body and to provide locomotion of the body with an alternating, reciprocal gait. Postural balance must be maintained, and, depending on one's position, few or many of these muscles act reflexly to maintain balance. The action of these muscles when the trunk moves on fixed limbs differs from the action that occurs when the limbs move with the trunk fixed. When one is standing still, the primary lower extremity muscles that are acting are those of the calf.

Plate 8.
Shoulder and Hip Joints

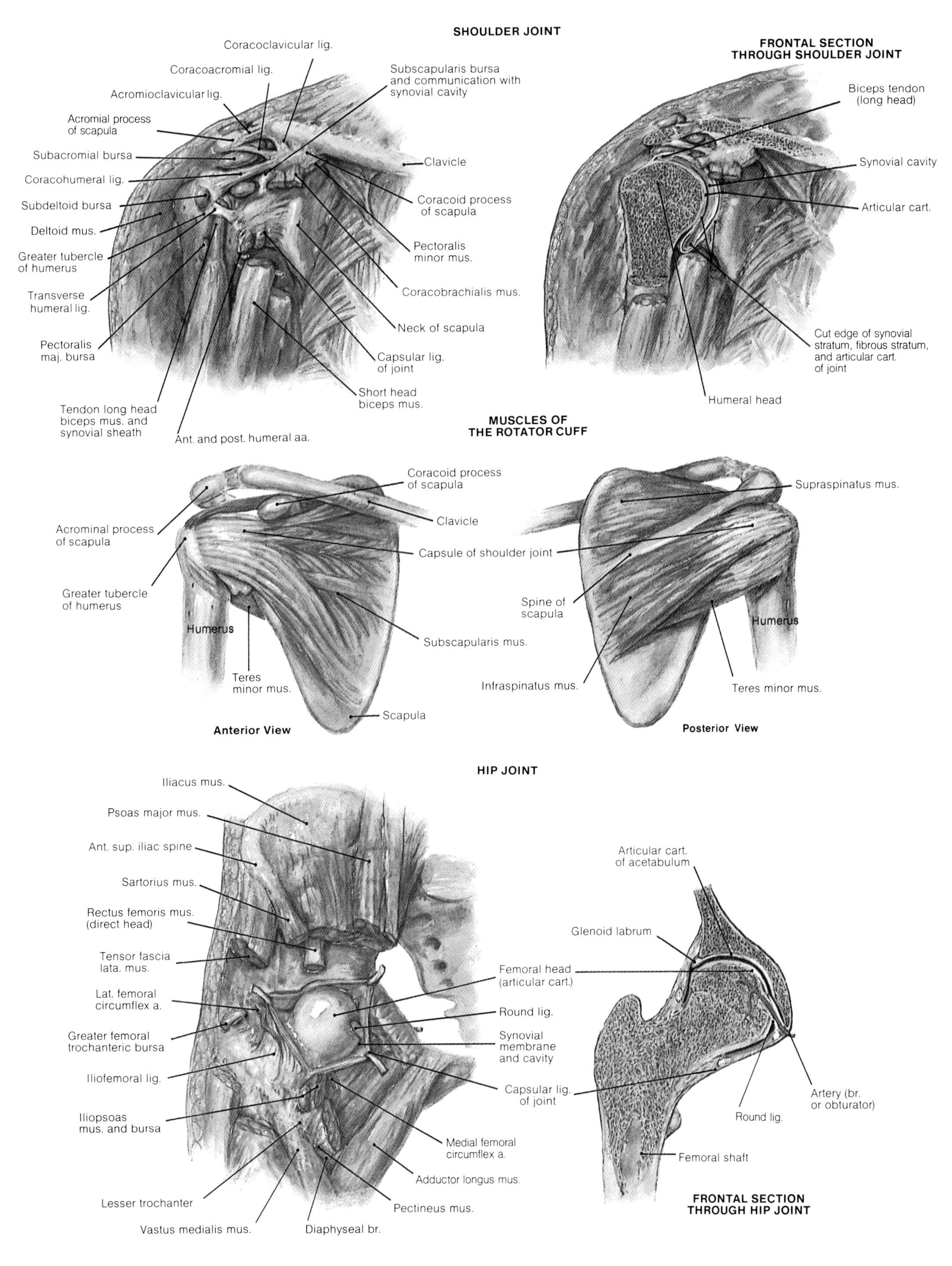

The hip is a ball-and-socket joint. Because it is required to bear the body weight, the hip has a deep socket, a strong capsule, and muscular attachments at a distance from the movement axis—all different from comparable structures in the ball-and-socket joint at the shoulder. The labrum adds further stability at the lip of the acetabulum. The capsule of the hip is cylindrical and has obliquely running fibers. The hip capsule has the least capacity in an internal rotation position. With synovitis or inflammation of the hip joint, internal rotation motion is lost. A leg held externally rotated and fixed, the position in which the capsule has the largest capacity, generally indicates that hip inflammation or injury is present. Two thirds of the femoral neck is intracapsular, and with infection of the proximal femur, a secondary joint infection may occur from seepage through the bone. Another unique aspect of the hip is the arterial supply to the femoral head; the medial and lateral femoral circumflex arteries run through the capsule from distal to proximal, in this way making them prone to injury from surgery or trauma. The principal bursae around the hip are the greater trochanteric bursa, beneath the gluteus maximus tendon, and the lesser trochanteric bursa, beneath the psoas tendon.

The muscular structures around the hip are designed for strength and power, not finesse. Flexion of the hip is provided by the iliopsoas, with the iliacus arising in the pelvis and the psoas arising from the anterior lumbar spine. The rectus femoris and the sartorius cross both the hip and the knee joints and also act as hip flexors. The gluteus maximus, the heaviest muscle in the body, provides hip extension. This muscle is best developed in human beings because of their erect posture, and it is this muscle that forms the buttock shape. Hip abduction is mediated by the gluteus medius and gluteus minimus muscles. Weakness of these hip abductors through injury or disease will cause a waddling gait. External rotators of the hip include the piriformis, obturator internus, superior and inferior gemelli, and quadratus femoris muscles.

There are three primary groups of thigh muscles. The posterior group, the hamstrings, are innervated by the sciatic nerve. The medial hamstrings are the semitendinous and semimembraneous muscles. The lateral hamstring is the biceps femoris. Because of the relative inelasticity of the hamstrings, originating on the ischial portion of the pelvis and attaching to the tibia, these muscles are a frequent source of injury in sports. The medial group, the adductors, are innervated by the obturator nerve. These muscles include the gracilis, adductor longus, adductor brevis, adductor magnus, pectineus, and obturator externus. The anterior group, the sartorius and the quadriceps femoris, are innervated by the femoral nerve and act as knee extensors. The sartorius is the longest muscle in the body. The quadriceps femoris is divided into the rectus femoris, which originates above the hip, and the vastus medialis, vastus intermedius, and vastus lateralis muscles. All the muscles of the quadriceps femoris coalesce distally to form the patellar ligament.

The knee, the largest joint in the body, is basically a modified hinge joint. Articulations are present between the tibia and the femur and between the patella and the femur. The patella is a sesamoid bone that is fully incorporated into the patellar tendon; it acts as a fulcrum to improve the quadriceps power during knee extension. The knee capsule is formed by a fibrous tissue extension from the muscle, and, because of its proximal muscle attachment, this capsule allows knee flexibility as well as stability. Four principal ligaments are present in the knee: medial and lateral collateral, and anterior and posterior cruciate ligaments. The lateral collateral ligament is closely associated with the iliotibial band, which is tightened proximally by the tensor fascia femoris muscle. The cruciate ligaments prevent anterior and posterior displacement of the tibia on the femur. The two menisci, wedge shaped in cross section, allow more even distribution of weight over the articular cartilage surfaces at the knee joint and may aid in lubrication of the articular surfaces. The lateral meniscus is O-shaped and the medial meniscus is a C-shaped structure. The medial meniscus is frequently torn in conjunction with medial collateral ligament injuries in sports. Because the menisci are avascular, a tear of these structures generally necessitates surgical removal because the scar tissue that forms remains weak and liable to reinjury. Medial rotation of the knee is effected by the popliteus muscle and by the sartorious, gracilis, and semitendinous muscles, which form the pes anserinus tendon at its insertion on the proximal medial tibia.

The fascia lata in the thigh, overlying the thigh muscles, is thin medially but thick laterally, forming the iliotibial band. Its proximal attachment is the tensor fascia femoris muscle, while its insertion aids in lateral knee stability. The lateral intermuscular septum extends to the linea aspera of the entire femur and is attached to the fascia lata, while the medial intermuscular septum is present principally only in the distal third of the femur. The primary bursae in the area of the knee are the prepatella bursa and the infrapatella bursa, although the pes anserinus tendon frequently has a bursa beneath it.

Below the knee, four compartments are formed in the leg by the deep fascial attachments. The anterior compartment is composed of the tibialis anterior, the extensor digitorum longus, and the extensor hallucis longus muscles. The lateral compartment contains the peroneus longus and the peroneus brevis muscles. These extensor muscles of the anterior and lateral compartments are innervated by the peroneal nerve. Posterior to the tibia and fibula, the deep posterior compartment consists of the tibialis posterior, flexor hallucis longus, and flexor digitorum communis muscles; the posterior

Plate 9.
Knee Joint

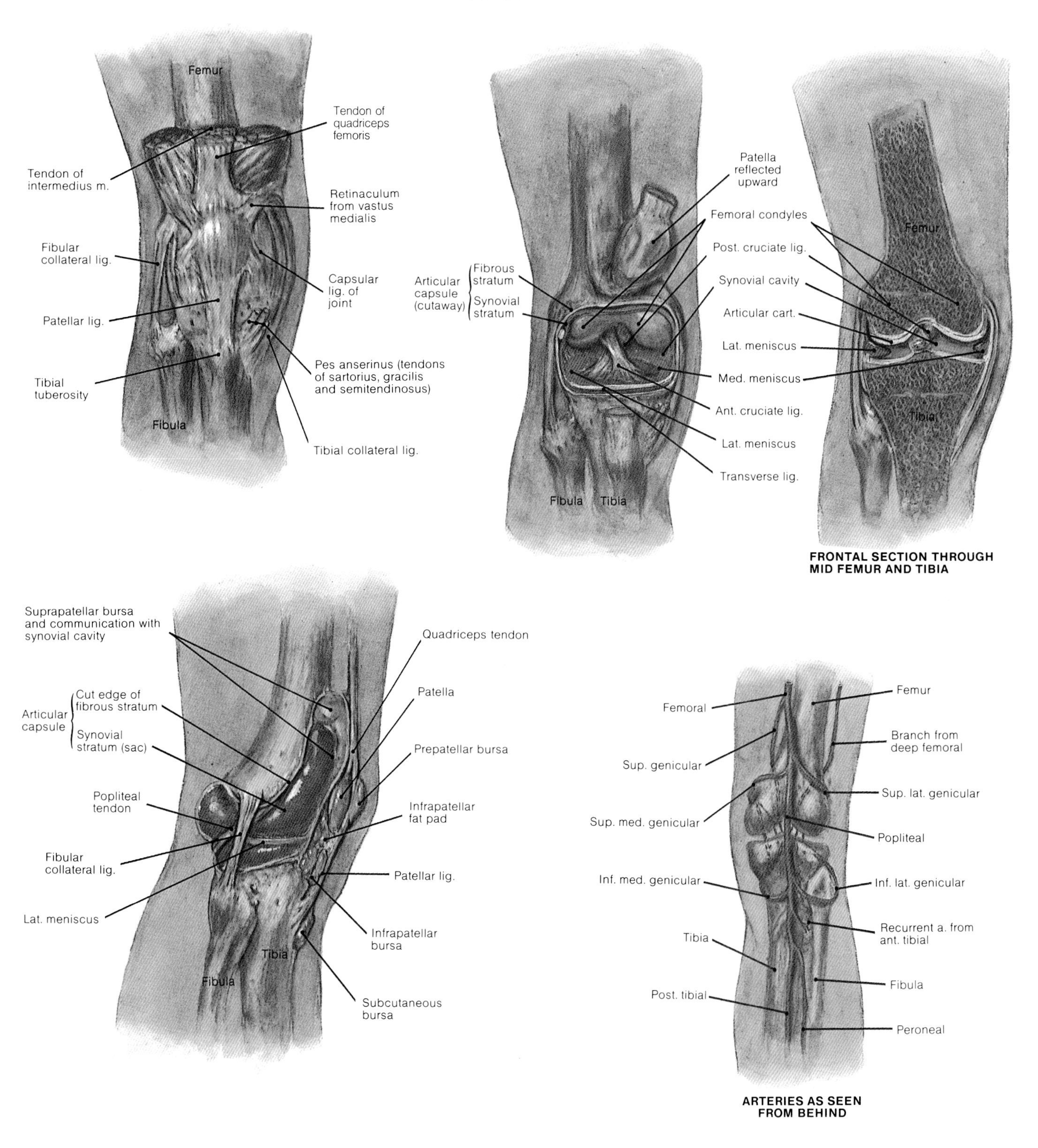

compartment contains the soleus, gastrocnemius, and plantaris muscles. These two compartments are innervated by the posterior tibial nerve. The flexor hallucis longus is a principal muscle for push-off in running or walking. The gastrocnemius and soleus compose the triceps surae and join distally to attach to the calcaneus as the Achilles tendon, the strongest tendon in the body. The gastrocnemius muscle originates above the knee, whereas the soleus originates below the knee.

The ankle joint is a hinge joint and functions to transfer weight through the tibia to the talus. The lateral collateral ligaments of the ankle are the anterior talofibular, calcaneofibular, and posterior talofibular; the major medial ligament is the deltoid. These ligaments are injured in ankle sprains with sudden inversion or eversion of the foot.

The function of the foot is to support and propel the body. The foot can adapt itself to different surfaces and absorb mechanical shocks well. The mobile medial side of the foot is more concerned with forward movement, while the lateral side of the foot functions primarily for static body support. The multiple bones of the foot are held in proper position by ligaments, muscles, and fascial bands. The plantar ligaments, primarily the spring ligament, support the arch and are supplemented in this function by the posterior tibial tendon and the plantar aponeurosis. The combination of these and other ligaments, muscles, and fascial bands provides a medial longitudinal arch. Imbalance of the muscles of the foot, as a result of neuromuscular disease or injury, causes foot deformity, because a muscle system that is usually so finely balanced cannot adapt to asymmetrical muscle activity. The intrinsic muscles of the foot include one dorsal group of short toe extensors and four layers of plantar, or sole, muscles. The first plantar layer includes adductor hallucis, flexor digitorum brevis, and abductor digiti minimi muscles. The second group contains the lumbricals, and the third is composed of the flexor hallucis brevis, adductor hallucis, and flexor digitorum minimus brevis muscles. The fourth group is the interossei muscles. These groups are covered by the plantar fascial aponeurosis superficially. The primary bursae of the feet are the retrocalcaneal bursa, deep to the Achilles tendon, and the intercalcaneal bursa, deep to the plantar aponeurosis. Bursae that form over the head of the first metatarsal or over the lateral head of the fifth metatarsal are called bunions.

Muscles, ligaments, and fascial structures and joints must work in close conjunction to allow smooth functioning of any body movement. Any abnormality in these structures, all derived from the embryonic mesoderm, will lead to limitation in joint mobility and compromise of normal body function.

The Abdominal Wall, the Inguinal Region, and Hernias

John J. White, M.D.

The abdominal wall is made up of the rectus abdominal muscles, which extend from the costal margins to the pubis on either side of the midline anteriorly, and the flat external and internal oblique and the transversus abdominis muscles lying laterally on each side. The rectus abdominis muscles are joined together by the strong fascia of the linea alba in the midline. The lateral muscles on each side originate from the costal cartilages superiorly, the lumbodorsal fascia posteriorly, and the iliac crests inferiorly. As they course anteromedially, they fuse together as a strong fascial aponeurosis, which attaches to the lateral margins of the rectus abdominis muscles. This lateral aponeurosis splits to provide both an anterior and posterior fascial sheath for the rectus abdominis muscle extending to the linea alba. In roughly the lower third, below the semilunar line of Douglas (Plate 10, dotted blue line), the aponeurotic fascia of the lateral muscles provides an anterior fascial sheath for the rectus abdominis, leaving only the endoabdominal fascia (transversalis fascia) posteriorly.

Thus, the peritoneal cavity is surrounded and shaped by a relatively strong muscular, aponeurotic, and fascial wall. Weak spots, however, may occur where muscle or fascia fail to fuse. Protrusion of an abdominal viscus through such a defect constitutes a hernia. If the viscus is not reducible back into the peritoneal cavity, it is termed an *incarcerated hernia,* and if its blood supply is compromised it is termed a *strangulated hernia.*

One area where a significant number of several kinds of hernias may develop is the groin, which comprises the inguinal and femoral canals and their contents. The inguinal canal is an oblique passage superior to the inguinal ligament (Poupart's ligament). Its entrance, the internal inguinal ring, lies just lateral to the inferior epigastric vessels, and it runs medially to the external inguinal ring, which lies over the pubic tubercle. The structures of the spermatic cord pass through the inguinal canal to the testis in the scrotum. These include the cremaster muscle and fascia, the vas deferens, the spermatic artery and veins, and the processus vaginalis. The ilionguinal nerve lies just under the aponeurosis of the external oblique in the canal and emerges through the external ring.

The inguinal canal is the site of direct and indirect inguinal hernias. Indirect inguinal hernias occur when a viscus, usually bowel, passes through the internal ring into the canal (Plate 11, white arrow), generally via the sac of a nonobliterated processus vaginalis. The vaginal process is a peritoneal outpouching of the abdominal cavity that proceeds through the inguinal canal into the scrotum, preceding, and perhaps guiding, the descent of the testis from its retroabdominal origin to its ultimate position in the scrotum. The vaginal process remains more or less patent in approximately 25 percent of men and therefore constitutes a potential sac for herniation throughout life.

Direct inguinal hernias, on the other hand, appear to be acquired on the basis of muscular weakness in the abdominal wall. The back wall of the inguinal canal is composed only of endoabdominal fascia (the transversalis fascia) and the peritoneum. This relatively weak area constitutes the inferior part of Hesselbach's triangle, whose borders are the lateral margin of the rectus abdominis muscle medially and the inferior epigastric vessels laterally. The inguinal ligament is its base. This weak back wall of the canal, in Hesselbach's triangle, is the site of origin of direct inguinal hernias (Plate 11, green arrow).

The femoral canal lies deep to the inguinal ligament and medial to the femoral vessels. Its medial margin is the iliopectineal ridge (Cooper's ligament) and lacunar ligament; its floor is the fascia over the pectineus muscle. This canal may be the site of femoral hernias

Plate 10.
Abdominal Wall—Inguinal Region

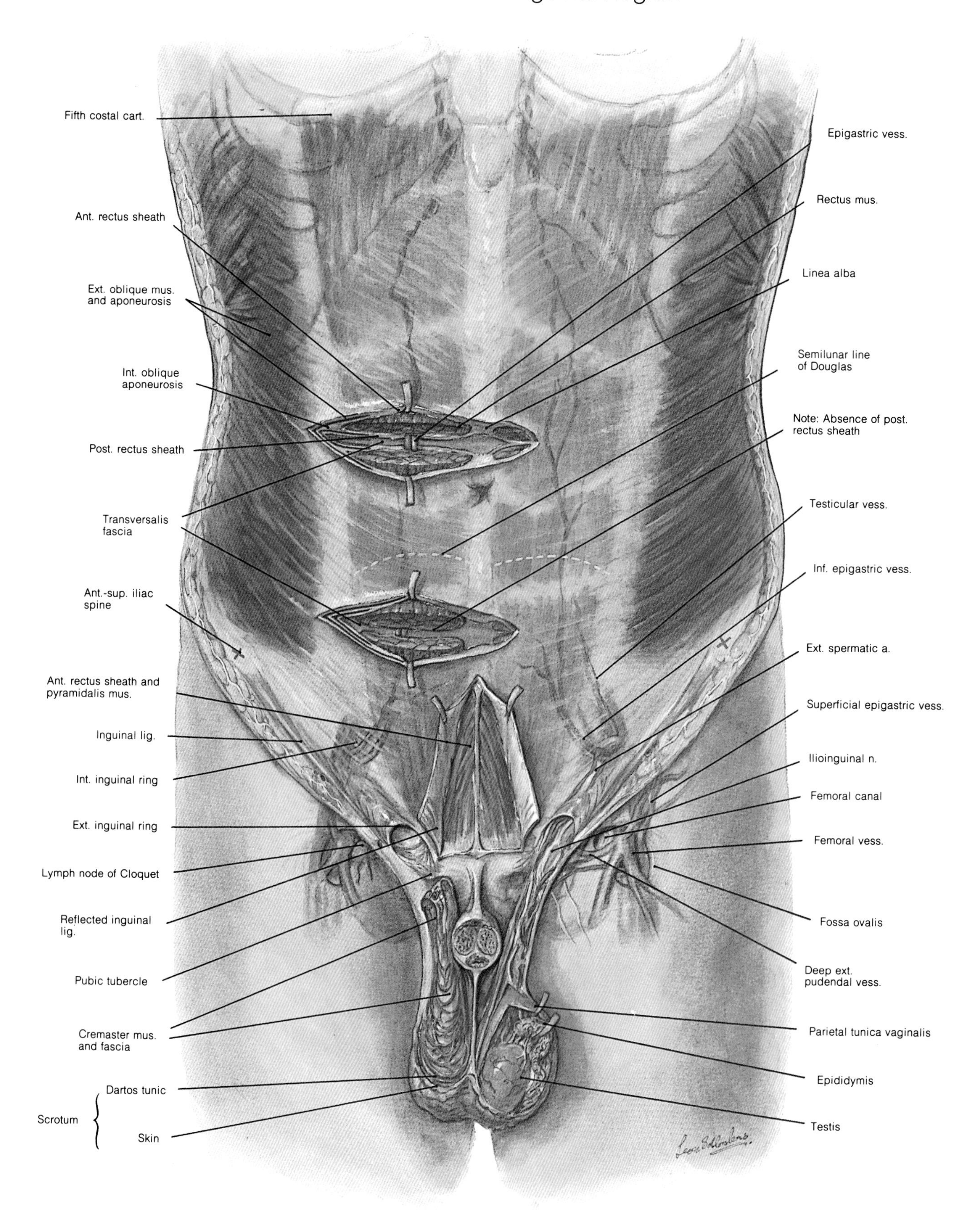

Plate 11.
Inguinal Region

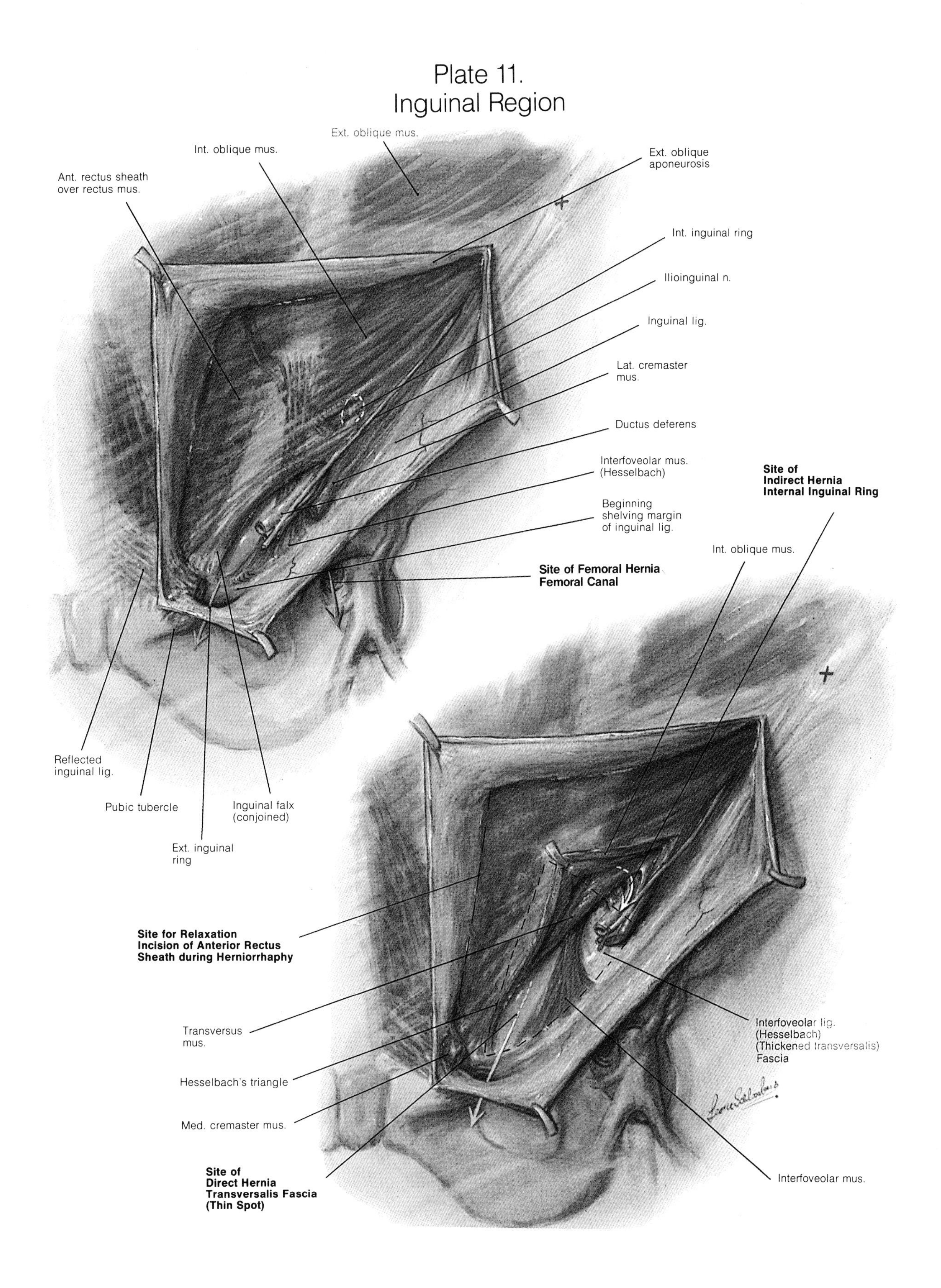

(Plate 11, pink arrow). Ordinarily it contains only fat and lymphatics that drain the leg, the groin, and the perineum.

The anatomical relationships of the common varieties of hernia are shown in Plate 12.

Indirect inguinal hernia is shown (upper left) in a completely patent processus vaginalis. There is dilatation of the proximal peritoneal sac within the inguinal canal due to visceral protrusion. The distal processus vaginalis in the scrotum about the testis is filled with fluid and waxes and wanes in size as the fluid leaves and enters the peritoneal cavity; this constitutes a communicating hydrocele.

A patent processus vaginalis is a more frequent occurrence than was once thought (upper right). Patency has been reported in 90 percent of neonates at autopsy and in 57 percent of babies dying before 1 year of age; an open processus vaginalis on the opposite side has been reported in 40 to 50 percent of both children and adults with unilateral inguinal hernias; and a patent processus vaginalis has been noted in 15 to 35 percent of adult autopsies. Indirect inguinal hernias in adults may occur into these open peritoneal sacs (arrow) as they do in children; in effect, they may remain as potential hernia sacs throughout life.

The distal processus vaginalis may also remain patent instead of obliterating. If the more proximal processus is obliterated, the tunica vaginalis of the testis may distend with secretions to form a noncommunicating hydrocele of the scrotum. An open area in the mid processus vaginalis can form an encysted hydrocele of the cord.

A double inguinal, or Pantaloon, hernia is shown at lower left. A typical indirect hernia emerges into the inguinal canal from the internal ring, along the same path as the patent processus vaginalis, and protrudes obliquely toward the external ring. A direct hernia is present as well, protruding through the weak back wall of the medial inguinal canal in Hesselbach's triangle. It bulges more directly toward the external ring. The interfoveolar muscle provides the support for the crotch of the pantaloons.

Some believe that most indirect hernias are congenital in nature, regardless of the age at which they occur, and that they consist of viscera entering the potential hernia sac represented by a patent processus vaginalis. Direct hernias generally appear to be acquired in nature, with peritoneal and visceral bulging through a weak point of the abdominal wall (the posterior inguinal canal).

A sliding inguinal hernia is shown at lower right. The peritoneal sac of an inguinal hernia may increase in size either by stretching of the peritoneum or by additional peritoneum sliding out of the abdominal cavity into the groin. With the latter mechanism, extraperitoneal viscera attached to the peritoneum may slide out as well. The sigmoid colon may present in this manner laterally in left-sided inguinal hernias (as pictured), or the cecum may present if the hernia is on the right side. The urinary bladder may present medially in cases of hernias on either side. Because of the sliding nature of these hernias, each of the viscera constitutes part of the respective wall of its hernia sac.

Plate 13 shows the sites of abdominal and pelvic hernias.

Hernias of the upper abdomen and diaphragm are shown in the upper left. Esophageal hiatal hernias occur through and about the esophageal hiatus of the diaphragm. This opening is most commonly formed by a decussation of the medial and lateral fibers of the right crus of the diaphragm, although it may be located within the fibers of the left crus in some instances. Two varieties of hernias about the hiatus are recognized.

The true hiatus hernia, or sliding hernia, is a displacement of the intra-abdominal portion of the esophagus, the cardiac sphincter, and perhaps the upper stomach into the chest through the hiatus. It is considered to be "sliding" because the stomach makes up part of the wall of the sac. Its symptomatology is different from that of the other types of hernias of the abdominal cavity, consisting not of incarceration and obstruction but of acid reflux, esophagitis, and possibly fibrosis.

The parahiatal hernia occurs less frequently and is a protrusion of stomach wall through the diaphragm near the hiatus, with a normally positioned esophagogastric junction. These hernias may incarcerate; esophagitis is not a problem, since the sphincter mechanism is intact.

The congenital posterolateral diaphragmatic hernia (Bochdalek's hernia) usually occurs on the left side and represents a persistently patent pleuroperitoneal canal. The diaphragm fails to form completely, and abdominal contents take residence in the chest as well. It is commonly associated with a hypoplastic lung on the side of the defect and constitutes one cause for acute respiratory distress in newborns. Treatment is directed toward reduction of the herniated abdominal viscera and repair of the defect.

The superior lumbar trigone lies beneath the latissimus dorsi muscle and is bounded superiorly by the twelfth rib, medially by the lateral margin of the paraspinous muscles (sacrospinalis), and inferiorly by the superior margin of the internal oblique muscle. Its floor is the aponeurosis of the transversus abdominis muscle. It may be the site of posterior abdominal wall hernias.

The inferior lumbar trigone (triangle of Petit) may also be the site for hernias. Its base is the iliac crest, and the floor the internal oblique muscle; the sides are the posterior margin of the external oblique and anterior border of latissimus dorsi just before they cross.

Hernias of the lower abdomen and pelvis are shown in the lower right. Umbilical hernias occur through a fascial defect in the linea alba at the umbilicus. This defect is usually congenital and frequently closes

Plate 12.
Hernias

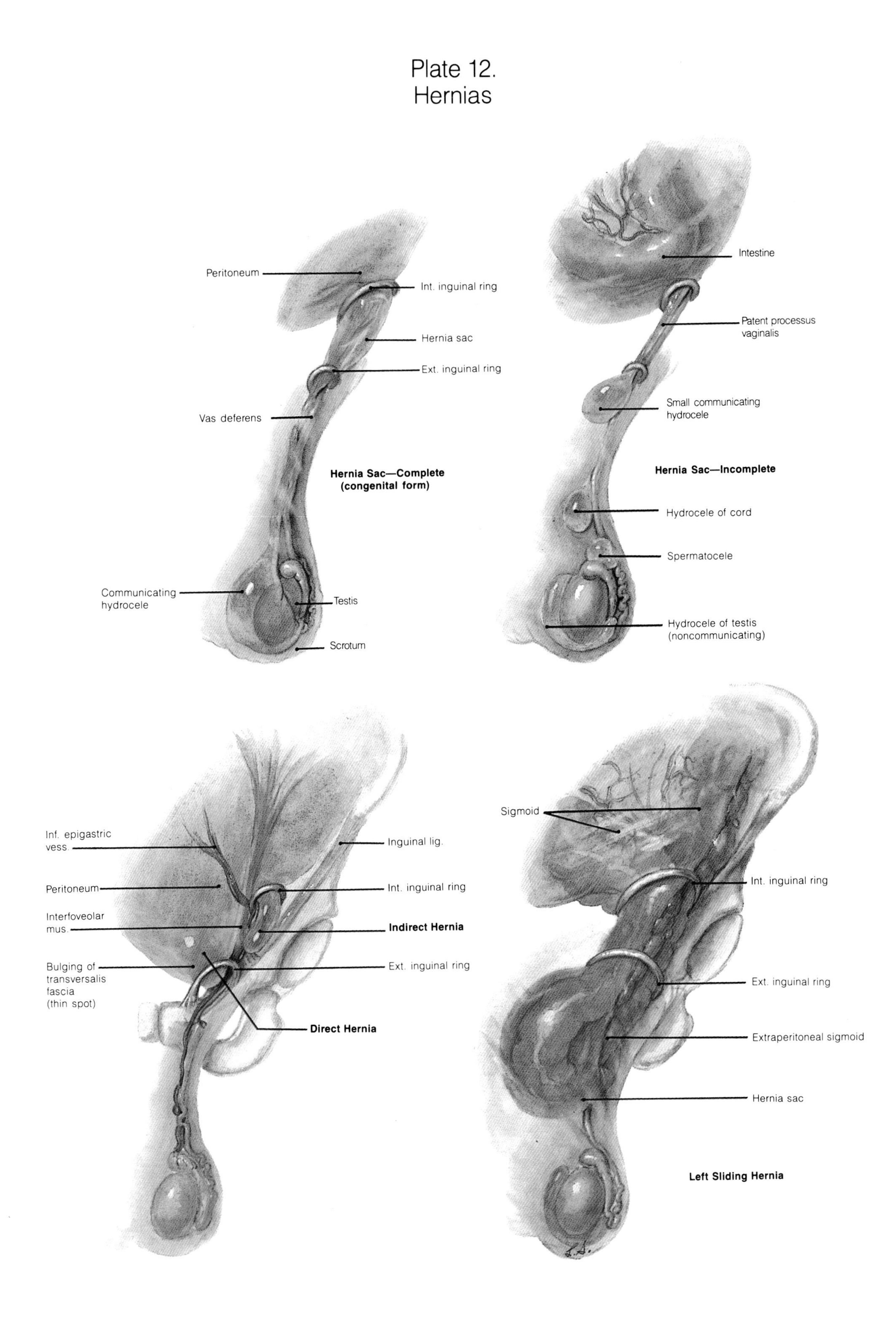

Plate 13.
Sites of Abdominal and Pelvic Hernias

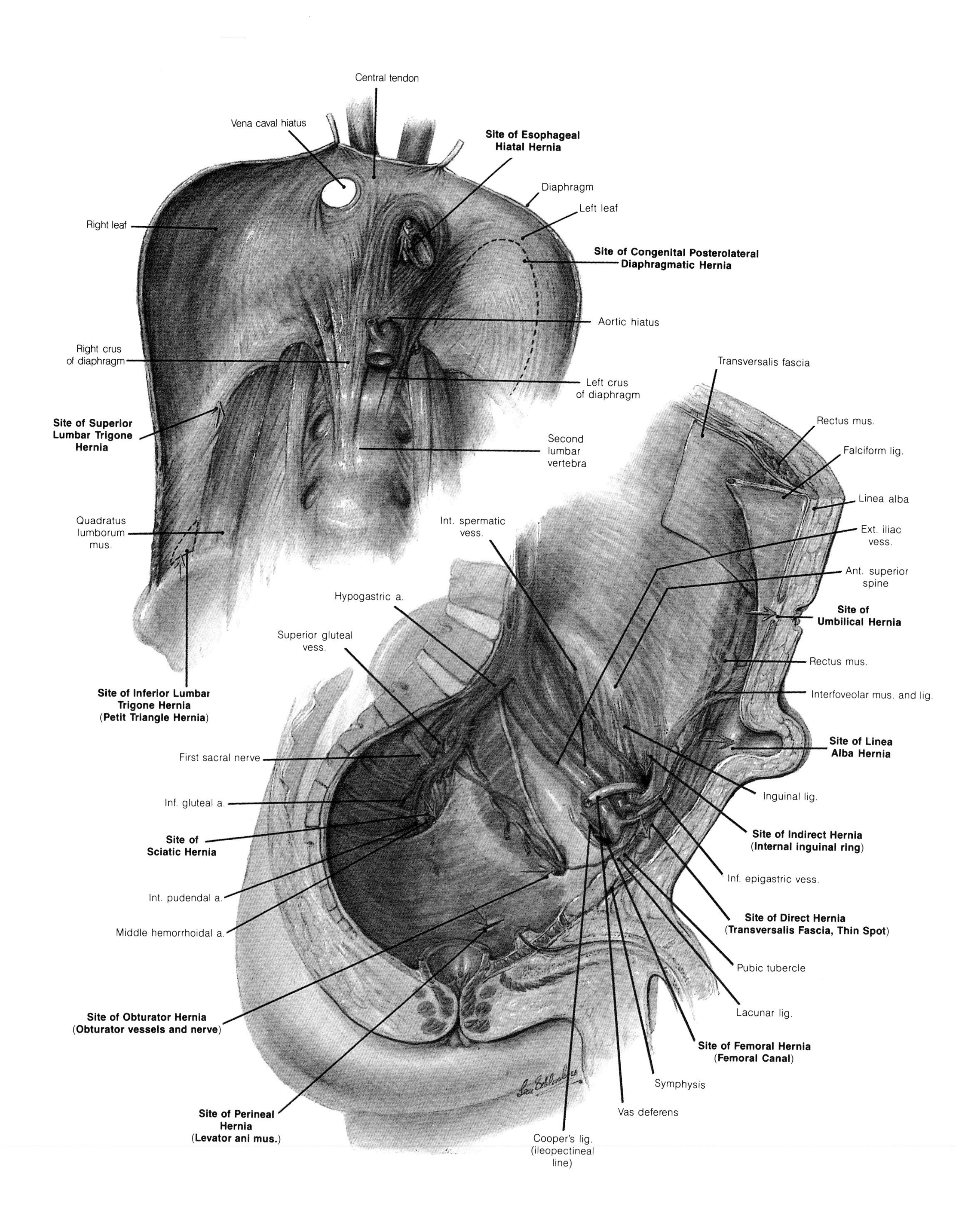

spontaneously. Defects larger than 1.5 cm in diameter in children tend to remain open, and repair is usually advocated in these cases. Incarcerated umbilical hernias are rare in children, but they occur more frequently in adults, especially parous women, and are associated with significant morbidity and mortality. Repair consists of reduction of sac contents and excision of the sac with fascia-to-fascia repair of the defect.

Hernias can develop anywhere along the linea alba. Preperitoneal fat can protrude through the orifices of penetrating vessels and nerves, with great pain because of pressure on the nerves. Adjacent peritoneum may also protrude, and abdominal contents may actually herniate in the fully developed state.

There are several types of pelvic hernias that occur rather infrequently. The sciatic hernia occurs through the great sacrosciatic foramen, the oval space between the innominate bone and the lesser sacrosciatic ligament through which the sciatic nerve and inferior gluteal artery pass. Obturator hernias occur as visceral protrusions through the obturator canal, through which the obturator vessels pass. Perineal hernias develop as viscera protrude through an opening, either natural or acquired, in the levator ani muscles which compromise the pelvic diaphragm.

The Hematopoietic System and Development of Blood Cells

5

William R. Bell, M.D.

The hematopoietic system is composed of red blood cells, white blood cells, and platelets, and their production sites and controlling sites responsible for cellular maturation and growth (e.g., stomach, liver), plus the fluid (plasma) in which these formed elements are suspended inside blood vessels (Plate 14). In the embryo, the source of these formed elements is the connective tissue called *mesenchyme.* In the human embryo, blood cells are first formed in the blood islands of the yolk sac. At a later time, when the embryo reaches 5 to 8 mm in length, the major source of blood cells is the liver, and a few weeks later production is supplemented by sites in the thymus and spleen. By the fifth month of gestation, production sites in the liver, thymus, and spleen gradually decrease and the bone marrow takes over hematopoietic production. The fixed mesenchymal cells are reduced to scant reticular stroma, but they remain throughout life with intact potentialities. Lymphocytes are an exception; they are produced in the lymph nodes, spleen, and thymus.

Erythropoiesis (production of red blood cells) in the infant and adult takes place continuously in the marrow of certain bones. The principal marrow sites are located in the skull, vertebrae, ribs, sternum, pelvis, femurs, and the humeri. As age progresses, the vertebrae, ribs, and sternum are the major sites of hematopoietic activity. Within the bone marrow, the red cell is derived from a primitive nucleated cell called the *erythroblast.* Proliferation results from successive mitotic cell divisions (Plate 15). As maturation progresses, hemoglobin appears and the nucleus becomes smaller and is eventually extruded from the cell.

The maturation process is a complex biochemical process regulated by numerous agents. Notable among these agents is an intrinsic factor produced by the stomach. The intrinsic factor, by complexing with an extrinsic factor (vitamin B_{12}), is responsible for its absorption from the intestinal tract into the blood. The mature red cell is then introduced into the circulating blood via the vascular channels of the bone marrow. In the adult there is 0.56 gm of marrow per gram of blood, and the bone marrow approximates 3 to 6 percent of the total body weight. A steady balance between red cell production and removal of senescent red cells (more than 120 days old) from the circulation by the spleen is accurately maintained. The rate of red cell production is normally controlled by a hormone called *erythropoietin,* which is mainly produced in the kidneys.

White blood cells (leukocytes) are independently motile cells, composed of three classes, each unique and different in morphologic structure and function: granulocytes, monocytes, and lymphocytes. In general, the leukocytes survive in the circulation for two to eight days. The most numerous of the leukocytes are the granulocytes, which originate in the bone marrow and can be divided into three subtypes: neutrophils, eosinophils, and basophils. Their orderly maturation and development from precursor blasts is shown in Plate 15. They are identified by a multilobed nucleus surrounded by numerous granules in the cytoplasm. Approximately 60 to 65 percent of the leukocytes in the body are neutrophils (pink cytoplasmic granules); and the eosinophils (red cytoplasmic granules) and basophils (dark blue cytoplasmic granules) total about 3 percent. Neutrophils function in defense and repair by performing phagocytosis of foreign cells, bacteria, and other infectious organisms. Eosinophils are phagocytic and participate mainly in antigen-antibody tissue interactions. Precise information on the function of the basophil is as yet lacking.

The largest cells in the circulating blood are monocytes, which total 7 percent of all leukocytes. Their origin is probably in the bone marrow. Monocytes are motile and are capable of phagocytosis. They are identified by a large, eccentrically placed, irregular nucleus, surrounded by a variable number of pink-purple and azurophilic granules.

Lymphocytes originate in the lymph nodes, spleen, thymus, and the tonsillar and lymphoid tissue of the alimentary tract and total about 25 to 30 percent of circulating leukocytes. Lymphocytes are identified by a single circular homogeneous nucleus that occupies most of the cell and is surrounded by a rim of cytoplasm that contains very few granules. Lymphocytes function in the body as the system responsible for acquired immunity to foreign cells and antigens. One type of lymphocyte is capable of producing immunoglobulins (antibodies), and the other type is concerned with cell-mediated immunity. The latter type is responsible for rejection of transplanted organs and certain allergic reactions.

Platelets (thrombocytes) are the smallest cells in the circulating blood. Like mature red cells they lack a nucleus and are not capable of cell division. Platelets originate as segmental structures that are released into the circulation from the cytoplasm of megakaryocytes, the largest cells in the bone marrow (Plate 15). The main function of the blood platelets is participation in hemostasis, the prevention and control of bleeding. In addition, platelets function in the maintenance of the integrity of the endothelial lining of vessels. Their circulation time in the blood is about ten days.

PLASMA

Plasma is a complex solution of electrolytes, proteins (7–8 percent), and water (90 percent). The major protein is albumin, but other proteins, including antibodies, hormones, lipids, and carbohydrate-protein complexes, and the various factors and components of the coagulation system, are present. The liver is the major site for the production of most of the proteins in the circulating blood. The liver is known to produce albumin, fibrinogen (Factor I), prothrombin (Factor II), and other coagulation factors that enable the blood to clot, including Factors V, VII, IX, and X. The liver also acts as a storage site for other agents that influence the production of elements in the blood. The blood produced by the hematopoietic system, the "milieu intérieur," is essential for normal function and life.

Plate 14.
Hematopoietic System

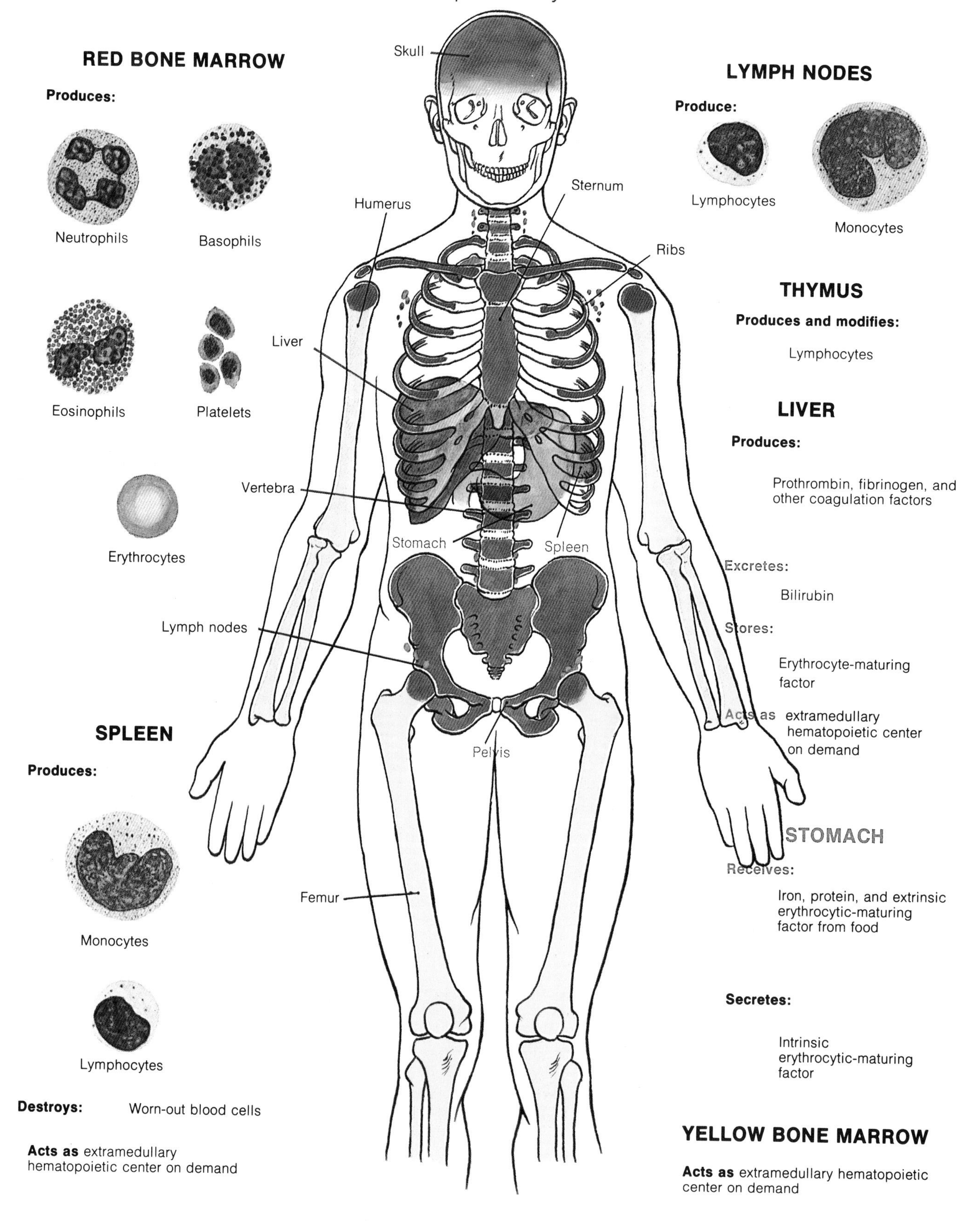

Plate 15.
Development of Blood Cells

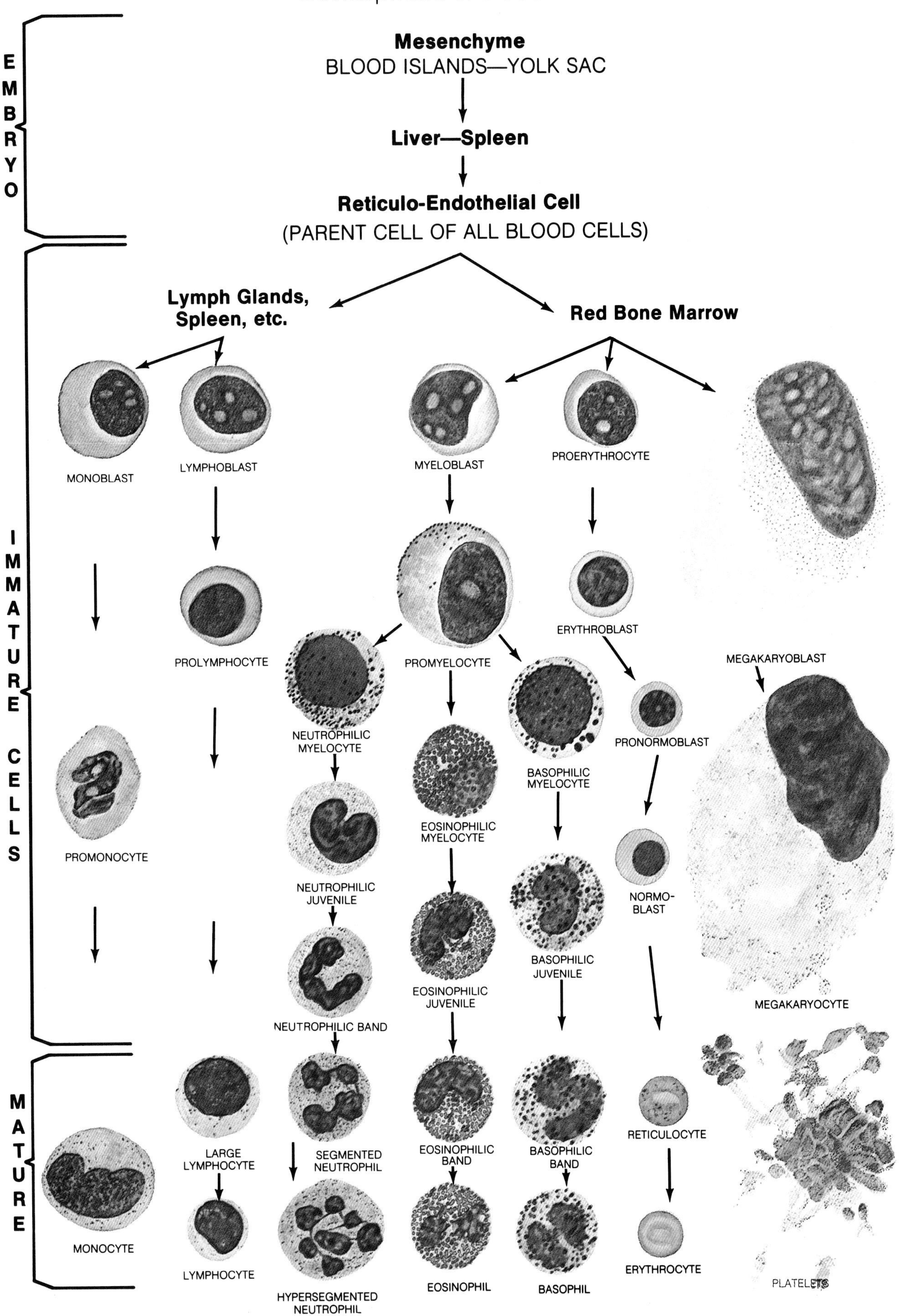

The Autonomic Nervous System

George B. Udvarhelyi, M.D.

The autonomic nervous system consists of a central and a peripheral component: information on central autonomic connections is still incomplete. A mechanism for interaction exists between the frontal cortex and the hypothalamus. The frontal cortex represents an afferent projection area that receives visceral impulses mainly from the hypothalamus, either directly or by way of the way stations in the thalamus. In response to stimulation, the frontal cortex activates other cortical areas, and issues efferent messages either directly to peripheral effectors or through the hypothalamus. Other cortical areas, like the cingulum and the posterior orbital, anterior insular, and temporal cortex, play important roles in cardiovascular function, respiratory movements, gastric motility, pupillary changes, and other autonomic responses such as piloerection, salivation, bladder contraction, and defecation. Through the cortical hypothalamic, corticothalamic, and corticostriate fibers, the different regions of the cerebral cortex are connected with the second important structure of the central part of the autonomic system, namely, the hypothalamus. Stimulation of various areas of the hypothalamus evokes special responses. Excitation of the anterior hypothalamic region produces bladder contraction, increase of gastrointestinal mobility, cardiac depression, and vasodilatation. Drowsiness, unconsciousness, and slowing of the heart occur in man after stimulation of the preoptic area. Excitation of the posterior and lateral regions of the hypothalamus results in the elevation of blood pressure, cardiac acceleration, pupillary dilatation, sweating, piloerection, hypoglycemia, and arrest of gastrointestinal movements. In animals, these effects are usually accompanied by expression of fear and rage. Control of body temperature is achieved in warm-blooded animals by release of excess heat through peripheral vasodilatation, sweating, and panting. Hunger and thirst are regulated by the hypothalamus, the appetite being inhibited by the ventromedial part, whereas the lateral and posterior hypothalamic regions seem to promote it. Stimulation of the dorsal hypothalamus increases the urge for drinking, whereas lesions of the hypothalamus result in hypodipsia. Observation in experimental animals and humans confirms that the anterior region of the hypothalamus is primarily concerned with the regulation of parasympathetic activities, whereas the posterior and lateral hypothalamic areas govern sympathetic responses. The pituitary gland is partly under the influence of the hypothalamus and partly controlled by a feedback mechanism of other endocrine glands. Through the anterior lobe of the pituitary gland, the hypothalamus participates in the control of the sex cycle, the activity of the adrenal cortex, and the function of the thyroid gland. Through the neuroendocrine connections to the posterior pituitary lobe, the hypothalamus controls the production of antidiuretic hormone and also the release of oxytocin.

The peripheral component of the autonomic nervous system consists of an efferent motor and an afferent sensory division. The peripheral motor autonomic division consists of two neurons: a preganglionic one, which has its cell of origin in the central nervous system, which synapses with several cell bodies of the second, the postganglionic neurons. The postganglionic nerve fibers terminate at the effectors, i.e., smooth muscle, glands, and heart. Preganglionic nerve fibers are covered with a thin myelin sheet and are white when viewed in the fresh state. Preganglionic neurons manufacture a chemical substance, acetylcholine, and are therefore called *cholinergic*. Postganglionic nerve fibers are unmyelinated and therefore are gray in appearance. Some postganglionic fibers are cholinergic, but the majority produce an adrenalinelike or noradrenalinelike substance for the stimulation of the effectors and are therefore called *adrenergic*.

According to the striking differences in the anatomical arrangement and functional significance, the motor autonomic nerves have been divided into two components, the parasympathetic division (craniosacral) and the sympathetic (thoracolumbar) outflow of the autonomic nervous system. The preganglionic parasympathetic nerve fibers leave the central nervous system at four places:

1. Hypothalamic outflow.
2. Tectal outflow. Preganglionic fibers originate from the Edinger-Westphal nucleus traveling along the third cranial nerve and terminate at the ciliary ganglion. The postganglionic fibers supply the sphincter of the pupil and the ciliary muscles. Stimulation produces contraction of the pupil and accommodation to near and far vision.
3. Bulbar outflow. Parasympathetic fibers arise from the nuclei of the rhombencephalon to innervate the salivary glands and the viscera in the thorax and abdomen. Preganglionic fibers travel along the facial nerve, with postganglionic participation in the lacrimal glands, and in the sphenopalatine, submaxillary, and sublingual glands. Stimulation produces increased secretion. The outflow of autonomic fibers along the facial nerve contains vasodilator nerve fibers for the middle meningeal arteries and vasomotor fibers for the blood vessels of the face. The glossopharyngeal nerve carries fibers to the parotid gland. The dorsal vagus nucleus is the origin of preganglionic fibers, with postganglionic release within the auricles of the heart and in the mesenteric plexuses of the intestinal wall. The vagal nerve fibers slow the heart, constrict the smooth muscle in the small bronchi of the lung, increase the activity of the pancreas and liver, and, except for the sphincter muscles of these regions, promote peristalsis of the stomach and intestine.
4. Sacral outflow. The preganglionic fibers arising from the lateral gray matter of the spinal cord travel with the ventrospinal nerve roots of the midsacral region into the pelvis. After synapsis in postganglionic cell bodies that are placed in the pelvic plexuses and in the walls of the bladder and rectum, the fibers terminate at the lower segment of the colon, the rectum, the bladder, and the genital system. Stimulation will produce contraction of the smooth muscles of the bladder and rectum and relaxation of the internal sphincters, resulting in emptying of the bladder and rectum.

The sympathetic division or the thoracolumbar outflow of the autonomic nervous system consists of short preganglionic neurons, which terminate with many collaterals at postganglionic cell bodies. The cell bodies of the postganglionic neurons, which have longer fibers, are remote from their organs of supply. They form conspicuous ganglia, which are placed along the side and in front of the spine, enabling sympathetic discharges to spread widely throughout the body, whereas parasympathetic impulses remain confined to more limited areas. The preganglionic sympathetic neurons are cholinergic, but the postganglionic sympathetic neurons by and large are adrenergic. The finely myelinated preganglionic fibers arising from the spinal cord travel along the ventral spinal roots of the thoracic and upper lumbar spinal nerves. Just beyond the intervertebral foramen, these fibers leave each of the spinal nerves as a white ramus communicans, which enters its corresponding paravertebral ganglion, which is present at each segment lateral to and along the entire length of the spine. Other portions of preganglionic fibers pass through and run as splanchnic nerves to the synaptic junctions in the prevertebral ganglia, such as the celiac, superior, and inferior mesenteric ganglia, which are placed in front of the spine. The unmyelinated postganglionic fibers arise from the paravertebral ganglion and join as gray rami communicantes the spinal nerves in which they travel to the blood vessels, sweat glands, and erector pilorum muscles of the skin, and probably also to the blood vessels of the striated muscles and bone. Stimulation will produce constriction of the blood vessels, increased sweating (hyperhidrosis), and piloerection. The regulation of body temperature is achieved by hyperactivity of the sweat glands and vasodilatation as response to heat, and by vasoconstriction and decreased sweat activity as response to cold, mediated through the sympathetic fibers alone. The term *sympathetic trunk* was coined to describe the chain or cord that extends from the base of the skull to the lower end of the spine.

The majority of afferent impulses from the viscera of the peritoneal cavity are transmitted by fibers of the sympathetic nervous system. These afferent fibers arise from the viscera and travel with the arteries of the viscera to reach a series of plexuses which are located around the major branches of the abdominal aorta. They traverse these plexuses without synapsis and become the splanchnic nerves. The afferent fibers subsequently enter the ganglia of the sympathetic chain, which they again traverse without synapsis to enter the spinal cord via the white rami. Within the spinal cord, the impulses traversing the viscerosensory nerve fibers stimulate the somatic sensory nerve fibers arising from the same level of the cord and innervating various dermatomes of the abdominal wall. Pain arising from an intraperitoneal viscus is therefore not localized to the involved organ but rather to the dermatome supplied by the peripheral nerve entering the cord at the same levels as the viscerosensory nerve fibers.

The celiac plexus is the largest of the autonomic plexuses and surrounds the celiac arterial trunk. The celiac plexus is connected to the thoracic sympathetic trunk via the greater, lesser, and least splanchnic

Plate 16.
Autonomic Nervous System and Effects of Stimulation (Respiratory and Digestive Systems)

Maintains Activity of Involuntary Visceral Effectors

(Organs below Level of Consciousness)

The hypothalamic nuclei and vasomotor and respiratory centers are principal sites for correlation and integration of impulses from cortex, visceral organs, and/or hormones in blood. Translate stimuli into physiological activity to maintain functional equilibrium.

EFFECTS OF STIMULATION
RESPIRATORY AND DIGESTIVE SYSTEMS

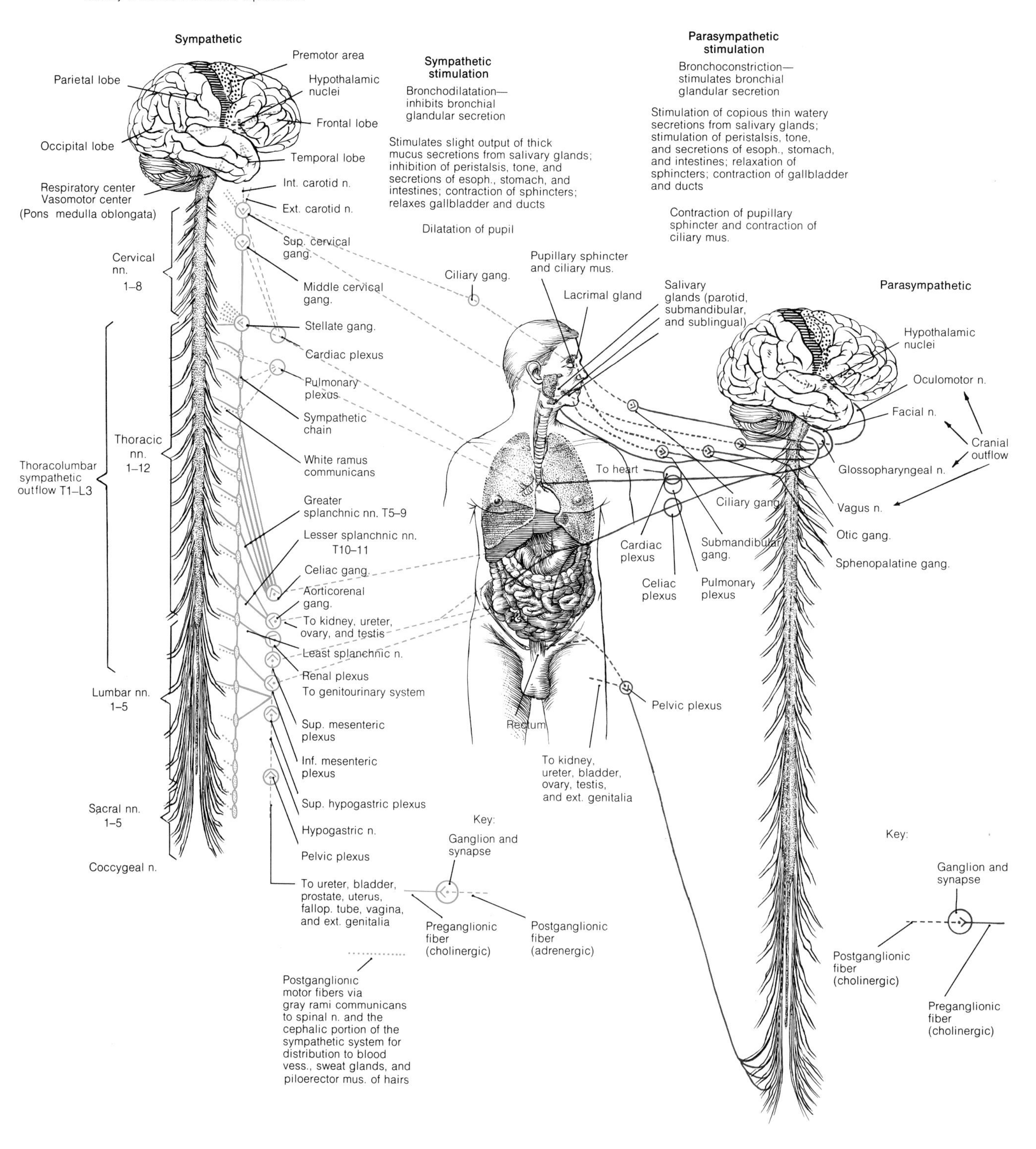

Plate 17. Autonomic Nervous System—Mechanism of Effector Organ Stimulation (Neurohormonal and Genitourinary Systems)

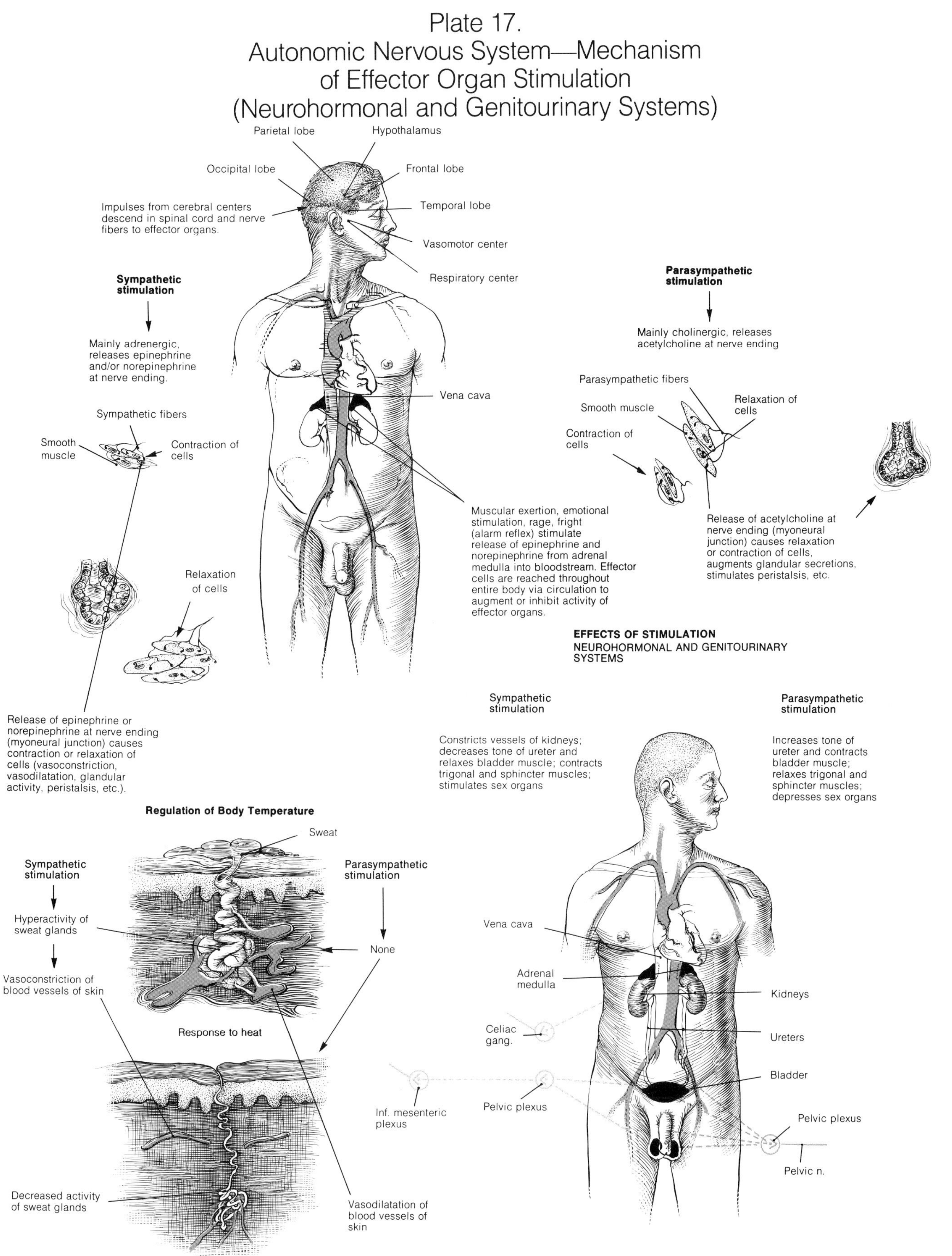

Plate 18.
Visceral Afferent Pathways and Referred Areas of Pain

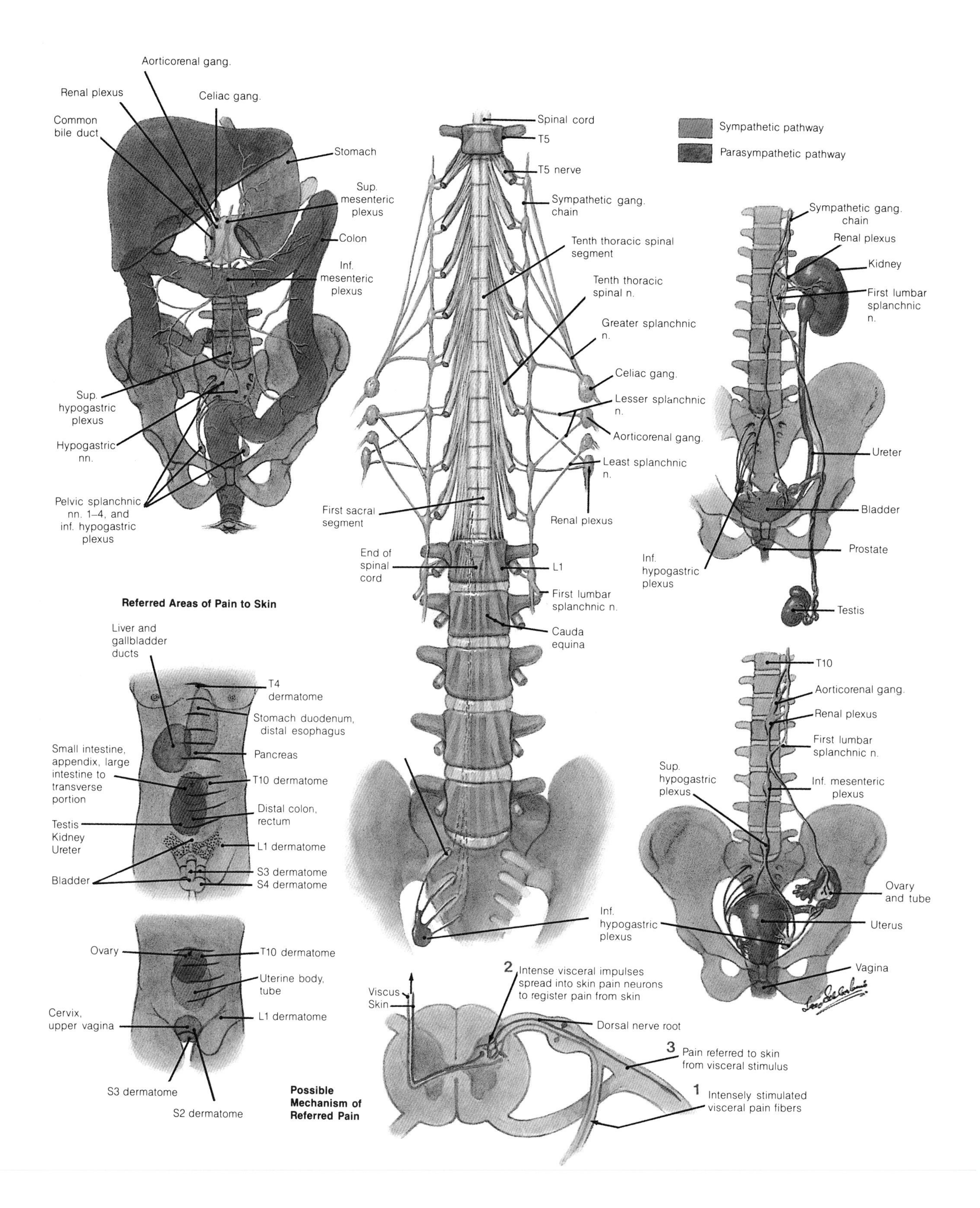

nerves. The greater splanchnic nerves enter the spinal cord from T5 through T9, the lesser splanchnic nerves enter from T10 to T11, and the least splanchnic nerves enter the spinal cord at T12. The celiac plexus transmits afferent sensory fibers from the stomach, liver, gallbladder, and pancreas which reach the spinal cord through the greater splanchnic nerve via the thoracic sympathetic chain. In addition, the celiac plexus also transmits afferent sensory fibers from the kidney, upper ureter, testis, and ovary which reach the spinal cord through the lesser splanchnic nerves via the thoracic sympathetic chain.

The superior mesenteric plexus receives afferent sensory nerves from the duodenum, entire small bowel, and proximal large bowel to the distal transverse colon. The plexus surrounds the origin of the superior mesenteric artery and is a continuation of the lower portion of the celiac plexus. The sensory nerves traverse this plexus without synapsis and travel via the lesser splanchnic nerves through the thoracic sympathetic chain to enter the spinal cord between T10 and T11.

The inferior mesenteric plexus is located around the origin of the inferior mesenteric artery and receives pain impulses from the distal part of the transverse and sigmoid colon. The impulses traverse this plexus, again without synapsis, and reach the lower thoracic and lumbar sympathetic chains via the least and first lumbar splanchnic nerves and enter the cord at T11, T12, and L1.

The superior and inferior hypogastric plexuses receive afferent sensory nerves from the pelvic organs and are located adjacent to the abdominal aorta and on either side of the pelvic viscera. They are joined to each other by the hypogastric nerves. Sensory fibers from the rectum, distal ureter, dome of the bladder, fundus of the uterus, and fallopian tubes traverse these plexuses and reach the spinal cord at the T11, T12, and L1 levels by traveling through the first lumbar splanchnic nerves and the higher lumbar and lower thoracic sympathetic chains. Pain impulses from the bladder neck, prostate, distal rectum, uterine cervix, and upper vagina are connected to the spinal cord through the inferior hypogastric plexus via the *parasympathetic pelvic splanchnic nerves* which enter the cord at the S2, S3, and S4 levels.

Painful stimuli, which give rise to impulses that pass along these afferent sympathetic and parasympathetic nerves, are triggered by three mechanisms: (1) an increase in intraluminal pressure of a hollow viscus; (2) ischemia of a viscus; or (3) the stretching of the capsule of a solid organ such as the liver, spleen, and kidney.

Abdominal pain can be divided into two phases. Changes in the character of the pain from one phase to the next indicate a progression of the disease process within the abdomen. The first phase of abdominal pain is called true visceral pain. It results from the distention of a viscus usually secondary to spasm of smooth muscle. Visceral pain is characterized as an ache that is not sharp or localized but diffuse and vague and generally referred to the midline. This pain is not associated with muscle rigidity or voluntary guarding of the abdominal muscles. The second phase of abdominal pain is called referred pain. This pain is due to continued stimulation of visceral pain fibers and subsequent involvement of both the visceral and somatic sensory nerve pathways. As a result of the crossover stimulation between the viscerosensory fibers and the sensory fibers of the somatic nerves, the pain is referred to the superficial area of the body surface or dermatome supplied by the peripheral somatic nerve arising from the same segment of the cord. This pain occurs when the irritable focus within a viscus has increased, giving rise to stronger stimuli, and usually indicates some type of pathology within the viscus. The pain, because of its relationship to the somatic dermatome, is fairly well localized to specific areas of the abdominal wall. This pain is usually associated with muscle guarding but not true rigidity.

It is important to appreciate the significance of these referred pain patterns, since they almost invariably indicate the presence of pathology within the abdominal cavity. Pain from the stomach and duodenum is referred to the epigastric area. Painful stimuli from the liver, gallbladder, and bowel ducts are localized to the right midepigastrium. Inflammation of the pancreas gives rise to referred pain in the epigastrium as well as directly through to the back. Pain from the small intestines is referred to the paraumbilical area. Pain from the appendix and large intestines up to the transverse colon is also referred to the paraumbilical area. Pain from the distal portion of the colon down to the upper rectum is referred to the suprapubic region. Pain from the kidney, upper ureter, ovary, and testicle is referred to the low epigastrium, paraumbilical, and flank areas. Bladder pain is referred to the suprapubic areas as well as the tip of the penis in males. Pain from the uterus, fallopian tubes, and ovaries is referred to the suprapubic area.

The physiologic significance of the autonomic nervous system has been summarized by Pick in the following way: The parasympathetic or cranial sacral component is essentially an anabolic system, because it is directed toward the preservation, accumulation, and storage of energies in the body. In contrast, the general effect of the sympathetic nervous system is catabolic because it causes the expenditure of bodily energies and inhibits the intake and assimilation of nutrient matter. There is, therefore, a high degree of stability of bodily function under the dual control of the autonomic nervous system. However, the importance of the ability of the body to cope with extremely difficult environmental conditions has been stressed and formulated by Cannon, who suggested a special designation, *homeostasis,* for the highly sophisticated coordination of physiologic processes that maintain most of the steady states in the organism.

7 The Anatomical Man

Leon Schlossberg

This figure displays many features that will benefit students at several levels of anatomical, medical, and surgical knowledge, as well as medical illustrators and physicians who wish to use it for reference.

The complexity of the anatomy is obvious, and portrayal of the systems and organs presented a challenge. A life-size illustration of the skeletal details was used as a background upon which the anatomical details were placed. Background sketches and radiopaque studies were used as well to portray living anatomy.

One of the objectives for this figure was to portray and highlight anatomical areas of clinical interest. An example is the heart: new medical and surgical techniques have given prominence to its general anatomy, with special emphasis on aspects of innervation relevant to the treatment of arrythmias and aspects of the blood supply relevant to operative procedures.

The phrenicocostal sinus is depicted at the level of the tenth rib. If one were to introduce a needle directed toward a hepatic duct, doing so at any higher level would likely involve the pleura, with the risk of pleural or pulmonary complications.

The portrayal of the pelvis highlights the prostate and its neurovascular supply as related to prostatic surgery and the maintenance of normal sexual function after radical prostatectomy.

The reader may find it useful to enlarge photographically any section of the figure representing the anatomy upon which an operative procedure is contemplated. The enlargement may then serve as a foundation upon which to plot the steps of the procedure on transparent overlays. It will be important to place two *X*'s on the foundation as registration points so that the overlays depicting successive steps of the procedure will align anatomically.

Plate 19.
The Anatomical Man

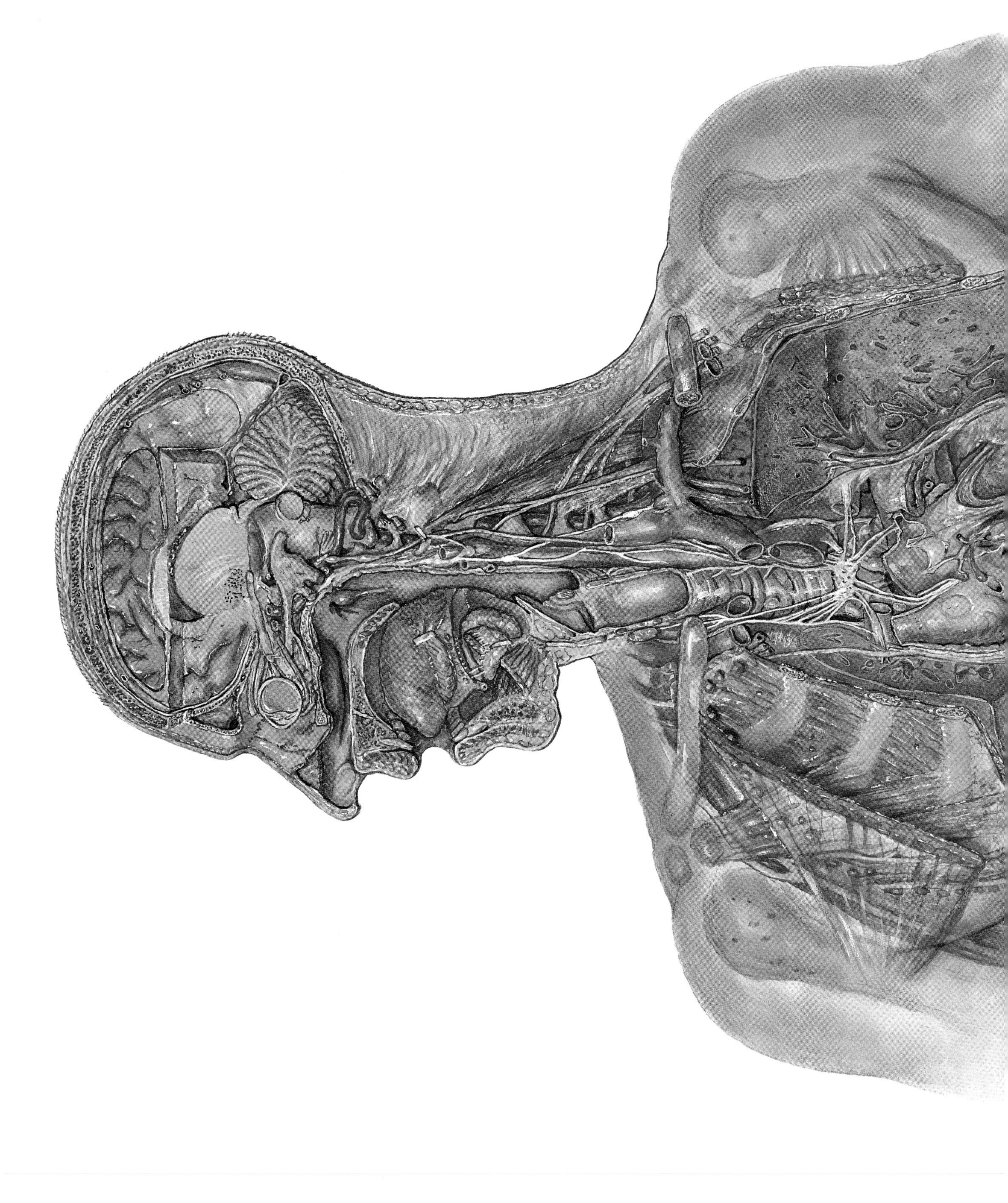

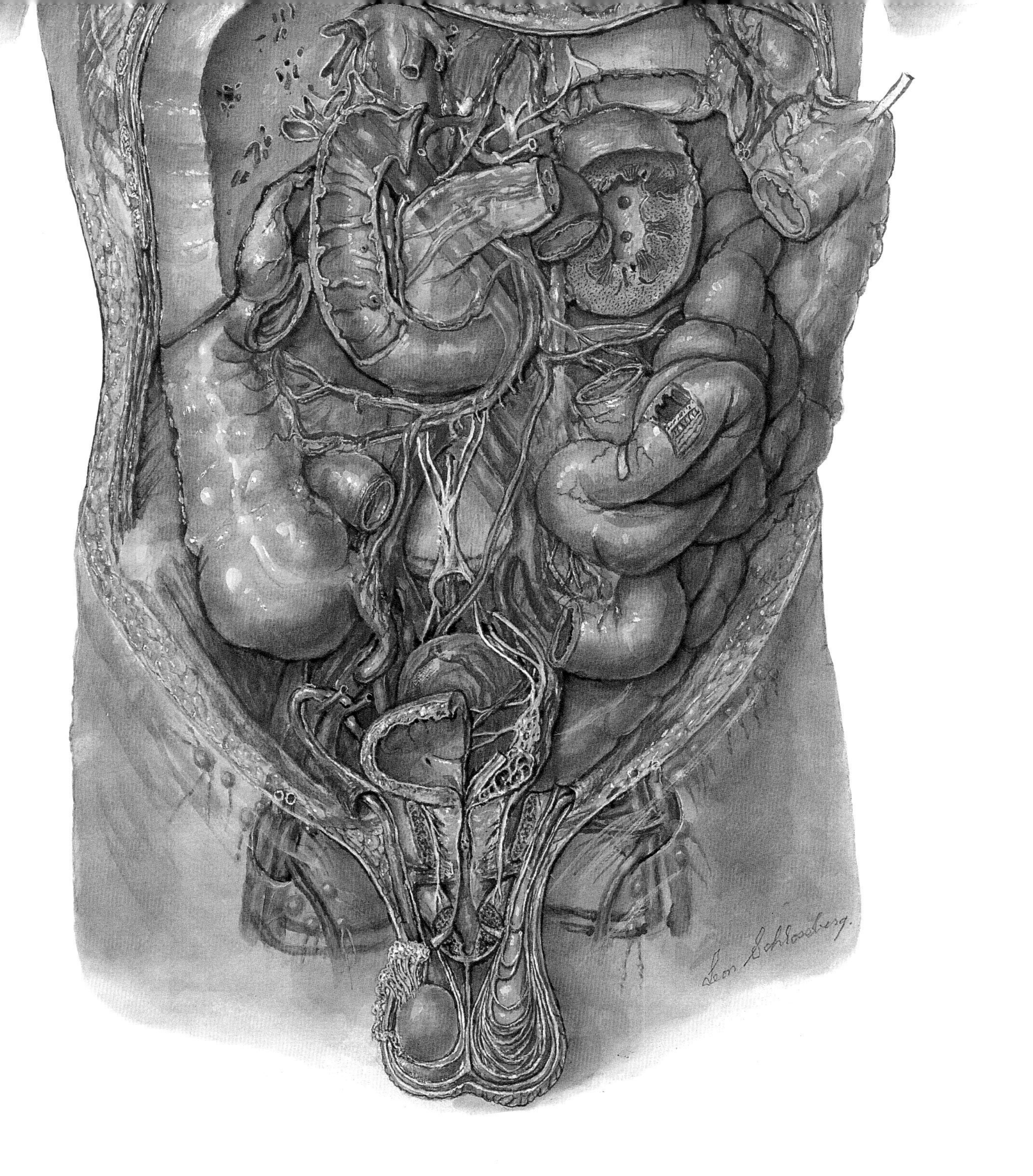
Leon Schlossberg.

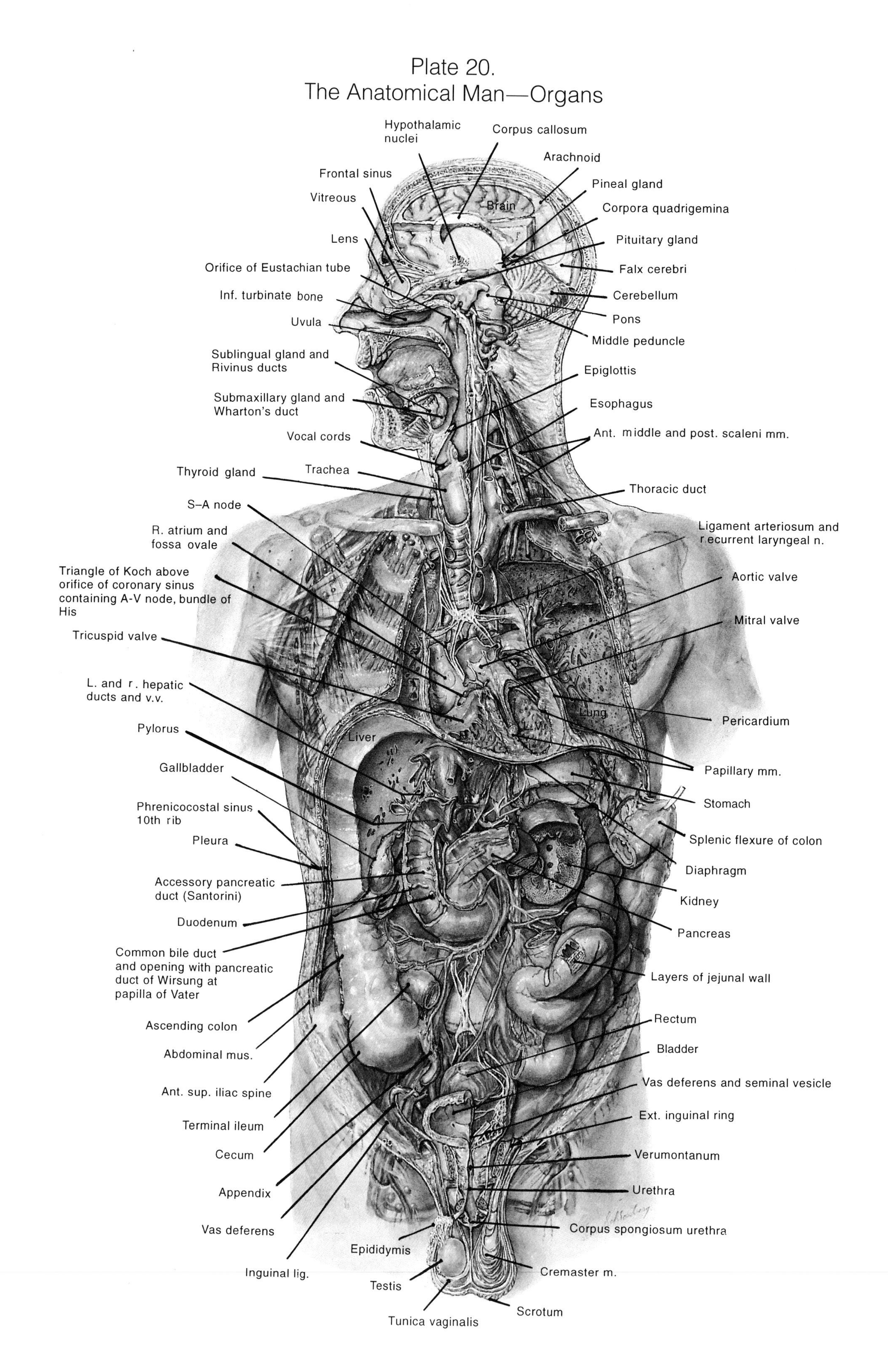

Plate 20.
The Anatomical Man—Organs

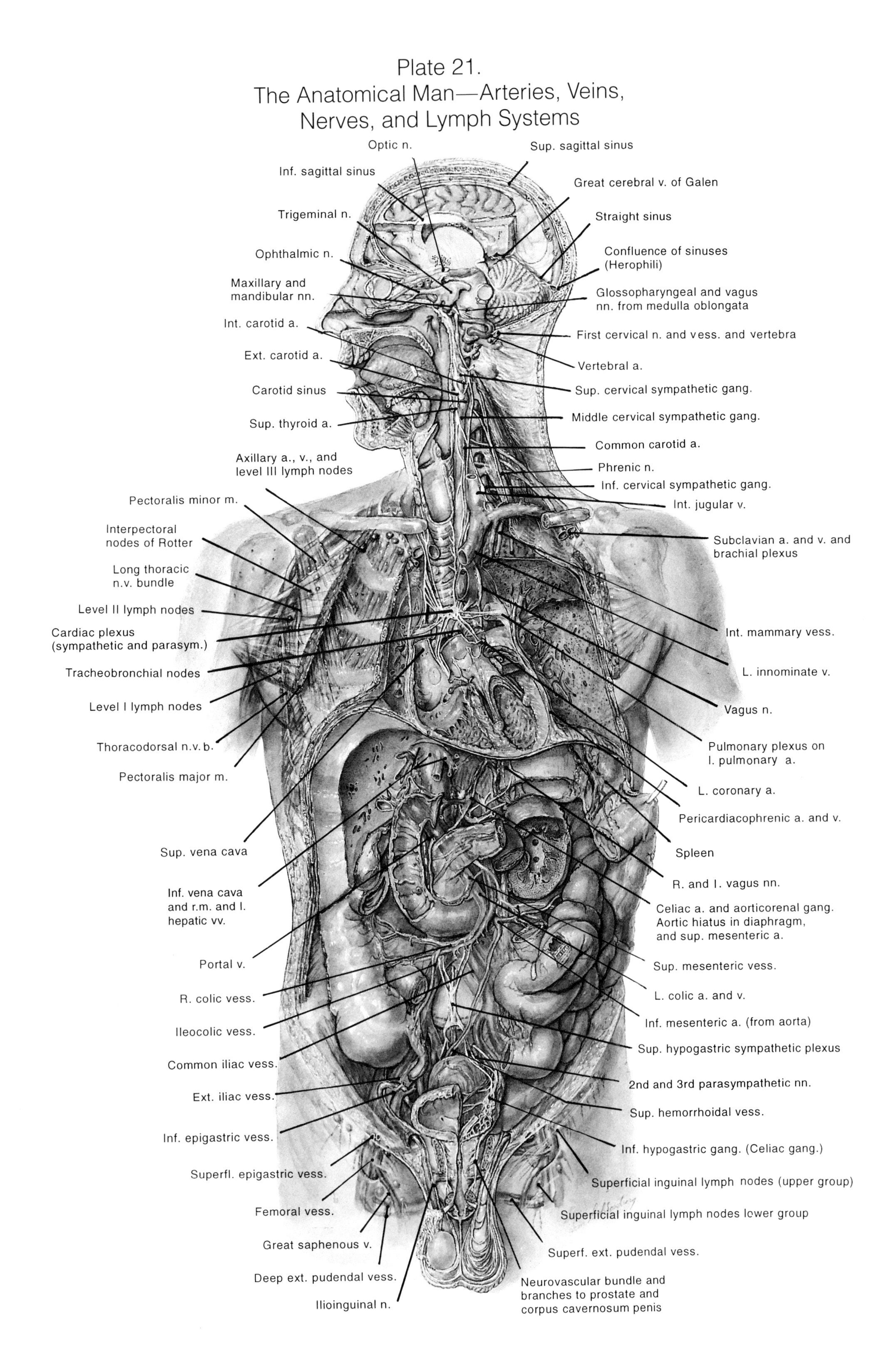

Plate 21.
The Anatomical Man—Arteries, Veins, Nerves, and Lymph Systems

The Aorta and Its Branches

Leon Schlossberg

The aorta is the body's main arterial blood vessel. Its branches carry blood from the heart to the entire body. Beginning with the two coronary artery branches, it supplies blood to the organs within the chest and abdomen and to the spinal cord, and ends with the iliac arteries, which supply blood to the legs. The wall of the aorta measures about 3 mm in thickness and progressively thins after branching. The wall is composed of an inner lining (intima), a middle and rather thick layer of muscle (media), and an outer layer (adventitia). Aging can change each layer, especially the intima.

Arteriosclerotic plaques may develop anywhere in the aorta and its branches but are more commonly found in the coronary arteries, the aorta itself, and its larger branches. These plaques are cholesterol-containing elevations on the inner surface, or intima, of the artery. They may present a problem when they become large enough to obstruct the flow of blood. On other occasions, the surfaces of plaques may become eroded and roughened. This produces an ulcerated lesion on the arterial wall and may result in a serious pathologic process in which clots form on the roughened surface and later embolize to block distal arterial branches.

Other conditions that may disturb the integrity of the arterial channel include dissection (separation of the layers of the vessel wall) and aneurysm (bulging of the arterial wall). The stretching or thinning of the wall may progress to rupture of the aneurysm, producing serious or fatal hemorrhage. Surgical operations have been devised to correct the pathologic conditions that may occur, from the excision of an aneurysm to replacement of the entire aorta and its large branches with prosthetic grafts.

The spinal cord receives segmental arteries at every vertebral level, and these vessels enter the spinal canal to supply the spinal cord. One especially large vessel that may come off the aorta anywhere from T7 to L4 is called the artery of Adamkiewicz. Interruption of this artery during the course of an operation often causes paraplegia. Techniques have been developed to preserve the blood supply to the spinal cord so that, during replacement of the aorta with a graft, paraplegia can usually be avoided.

This figure was prepared to be useful in teaching, and to illustrate to surgeons the normal anatomy to be reconstructed.

Plate 22.
Aorta and Its Branches—Lateral View

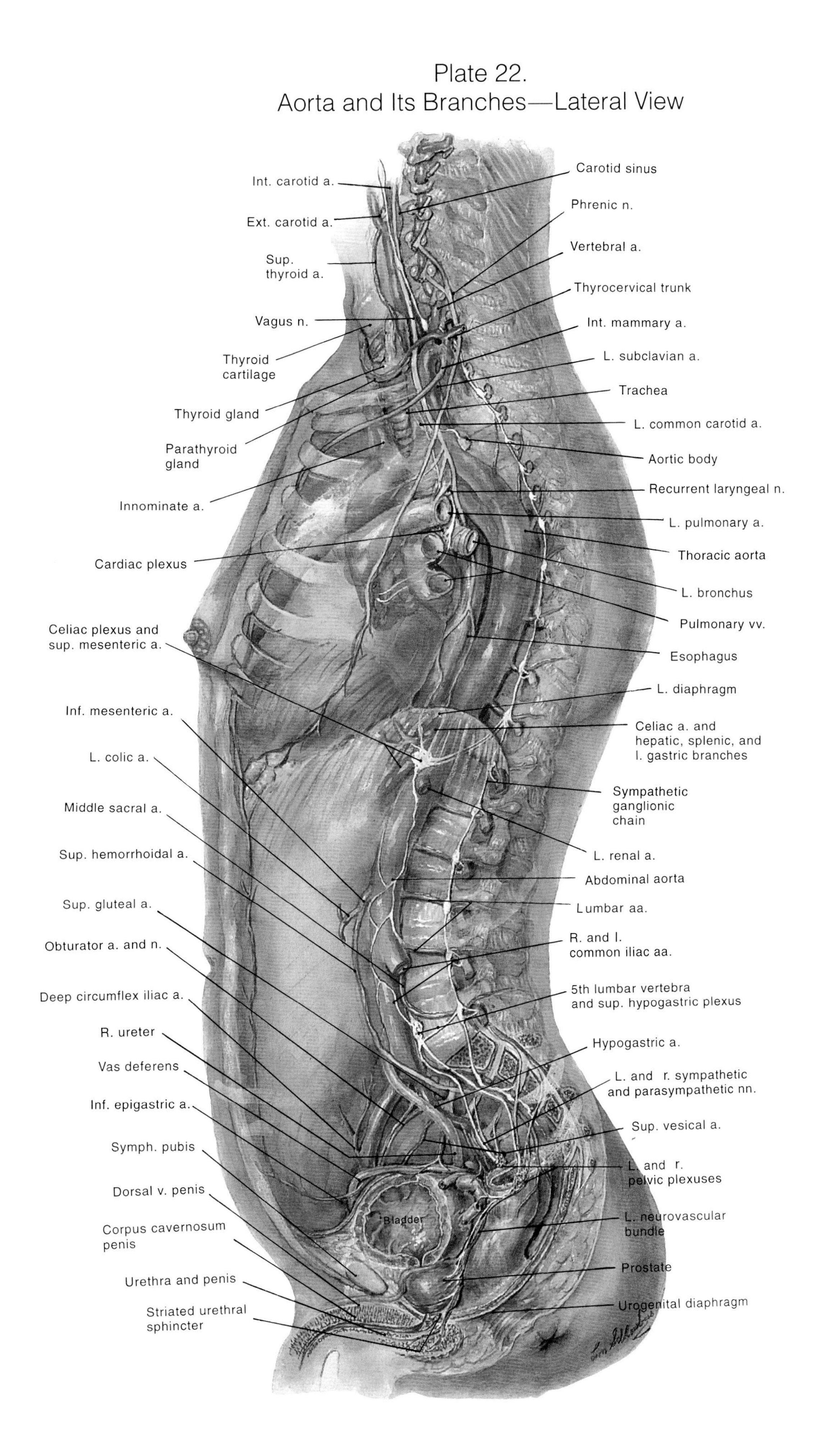

The Peripheral Nerves

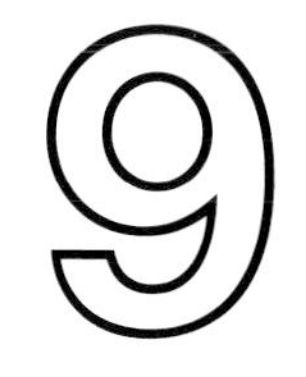

George B. Udvarhelyi, M.D.

The peripheral nervous system consists of thirty-one pairs of segmentally arranged spinal nerves that connect the spinal cord with the various parts of the body. There are eight cervical, twelve thoracic, five lumbar, five sacral, and usually one coccygeal spinal nerve. The first cervical nerve emerges between the occipital bone and the atlas; the eight cervical nerves emerge between the seven cervical and the first thoracic vertebrae; and below that, each spinal nerve emerges from the intervertebral foramen between its own and the next lower vertebra. Each spinal nerve has a posterior and an anterior (dorsal) afferent and ventral (efferent) root. At the level of the intervertebral foramen, the posterior root forms the spinal ganglion, which contains the central origin of the afferent fibers. The ventral root does not have a ganglion but joins the posterior root, and both emerge from the intervertebral foramen as a mixed spinal nerve. This common nerve trunk contains both afferent and efferent fibers. The ventral roots, with the exception of the cervical roots, contain efferent fibers of the sympathetic nervous system.

Shortly after the spinal nerves have been formed by the union of the two roots, they divide into ventral and dorsal primary divisions. The dorsal primary division splits into branches, which supply the dorsal-axial musculature and adjacent skin. The ventral primary divisions, with the exception of those in the thoracic region, form four main plexuses: cervical, brachial, lumbar, and sacral. Those in the thoracic region, except the first, remain separate and divide individually into branches that supply the muscles and skin of the thoracic and abdominal walls. In the formation of the four main plexuses, the nerve fibers rearrange themselves into peripheral nerves. In addition to the motor and sensory impulses carried by the posterior and anterior roots, the sympathetic nervous system participates as well, this participation consisting of visceral efferent and visceral afferent fibers, which control the vasomotor, visceral, and sweat gland function throughout the somatic areas.

From the anatomical description it is clear that almost all peripheral nerves are mixed nerves containing: (1) efferent or motor fibers, (2) afferent or sensory fibers, and (3) postganglionic fibers of the sympathetic nervous system. Most peripheral nerves are composed of fibers of three or more spinal segments. Certain cranial nerves (III, VII, IX, and X) and the sacral nerves of the second, third, and fourth segments contain preganglionic fibers of the parasympathetic autonomic nervous system.

Because peripheral nerves are mixed nerves, some anatomical principles should be emphasized that have considerable clinical importance. In terms of the motor component of the peripheral nerve, there is distinction between a radicular paralysis and a peripheral nerve paralysis. If there is paralysis of a group of muscles, each supplied by the same peripheral nerve, the site of the lesion must be peripheral. On the other hand, if there is paralysis of a group of muscles each of which has the same radicular innervation, the lesion must be located in the anterior roots or spinal cord.

The cervical plexus (C1–C4) innervates the deep cervical muscles, serving the function of flexion, extension, and rotation of the neck. The cervical nerves also innervate the scalene muscles, and the phrenic nerve the diaphragm, both serving inspiration.

The sensory branches supply the skin of the lateral occipital portion of the scalp, the upper median part of the auricle, and the area over the mastoid process innervated by the small occipital nerve (C2–3); the skin of the back of the ear and over the front of the neck innervated by the cervical cutaneous nerves (C2–3). The supraclavicular branches (C3–4) supply the skin over the clavicle and upper deltoid and pectoral regions as low as the third rib.

Irritative lesions like meningitis or high spinal cord tumors are occasionally responsible for symptoms of a cervico-occipital neuralgia; peripheral lesions are relatively rare in the cervical region because of the protection afforded by the surrounding muscles.

The brachial plexus (C5–T1) is formed by the anterior primary divisions of the last four cervical and first thoracic nerves. The relationship of the brachial plexus with the cervical-thoracic vertebral column, with the first rib, with Sibson's fascia, and with the three scalenus muscles plays an important part in the functional anatomy of this structure. As Plates 23 and 24 show, both the relationship of the plexus to these structures and also the various clinical symptoms can be better understood if one grasps the organization within the plexus itself. The brachial plexus is composed of three different components:

The roots of the plexus: C5–C6 fuse to form the upper trunk; C7 becomes the middle trunk and C8 and T1 fuse to form the lower trunk.

The middle portion of the plexus is formed by three trunks; each divides into anterior and posterior divisions. The lateral cord is formed by the anterior divisions of the upper and middle trunks; the medial cord is formed by the anterior division of the lower trunk, and the posterior cord is formed by the three posterior divisions of the upper, middle, and lower trunk. Finally, the three cords, lateral, posterior, and medial, split to form the main branches of the plexus. The position of the cords is in relation to the axillary artery. Branches from the medial and lateral cords form the median nerve; the rest of the lateral cord becomes the musculocutaneous nerve. Another portion of the medial cord forms the ulnar nerve; and the posterior cord divides to become the radial and axillary nerves. Compression by an anomalous cervical rib, by the first rib, and by the scaleni muscles will give rise to well-defined clinical symptomatology.

Numerous smaller nerves arise from the roots of the plexus: the posterior thoracic nerves (dorsal scapular nerve) (C5), the long thoracic nerve (of Bell) (C5–6–7), and rami to the scaleni and longus colli muscles. Several branches arise from the trunks: to the subclavius muscle and to the supra- and infraspinatus muscles by the suprascapular nerve. Several branches from the cords supply the pectoralis major and minor muscles through the medial and lateral anterior thoracic nerves; the upper subscapular nerve from the posterior cord innervates the subscapularis muscle (C5–6); the latissimus dorsi muscle is innervated by the middle (long) subscapular and thoracodorsal nerves (C7–8); and the lower subscapular nerve (C5–6) innervates the teres major and part of the subscapularis muscle. It should be mentioned that from the medial cord (C8–T1) sensory branches go to the medial surface of the forearm (medial antebrachial cutaneous nerve) and to the medial surface of the arm (medial brachial cutaneous nerve). The main branches of the brachial plexus are the musculocutaneous nerve (C4–5–6), the median nerve (C5–6–7–8 and T1), and the ulnar nerve (C8, T1). Nerves from the plexus trunks and partly from the cord innervate the muscles servicing the movements of the scapula, and elevation, rotation, abduction, and depression of the arm.

The musculocutaneous nerve innervates the biceps and coracobrachialis and brachialis muscles and controls flexion and supination of the forearm and elevation and abduction of the arm. The median nerve innervates the flexors of the hand and fingers (except the flexor carpi ulnaris and the ulnar half of the flexor digitorum profundus) and the pronators serving the flexion of the hand and fingers and pronation. It may be compressed by the thickened transverse ligament (Plate 24). The ulnar nerve innervates the ulnar flexor muscles, the adductor pollicis, hypothenar muscles, and third and fourth lumbricales and the interossei muscles, serving the movements of the little finger, flexion of the first phalanx and extension of the other phalanges of the fourth and fifth fingers, and the spreading apart and bringing together of the fingers. Its superficial location in the olecranon groove may give rise to dysfunction of compression, requiring a surgical transposition. The radial nerve provides the innervation of the triceps, brachioradialis, and extensor muscles of the fingers, serving extension, partial flexion, and supination of the forearm, hand, and fingers. The thoracic nerves innervate the thoracic and abdominal muscles, serving elevation of the ribs (expiration and abdominal compression).

The lumbar plexus is located in the substance of the psoas muscle and is the upper portion of the lumbosacral plexus. There are five pairs of lumbar nerves; the anterior primary divisions of the first three, part of the fourth, and, in 50 percent of the population, part of the twelfth thoracic nerve contribute to the lumbar plexus. Most of the branches of the lumbar plexus, which derive from the upper lumbar nerves, innervate the skin: the iliohypogastric nerve (T12, L1), the upper lateral part of the thigh and the symphysis; the ilioinguinal (L1), the upper medial part of the thigh and the root of the penis and scrotum; the genitofemoral nerve (L1–2), the scrotum and labia; the lateral femoral cutaneous nerve (L2–L3), the anterolateral side of the thigh. This nerve is responsible for unpleasant paresthesias (meralgia parasthetica) if it is strangulated or partially damaged.

The lumbar plexus (T12–L4) gives rise to the femoral nerve (iliopsoas, sartorius, and quadriceps muscles) and to the obturator nerve (the adductors of the thigh, gracilis, and external obturator muscles), serving flexion of the hip and upper and lower leg, extension of the lower leg (femoral), and adduction and outward rotation of the leg (obturator).

The sacral plexus: The five pairs of sacral nerves form the posterior primary division, which gives medial

Plate 23.
Peripheral Nerves
(Exclusive of Cranial Nerves)

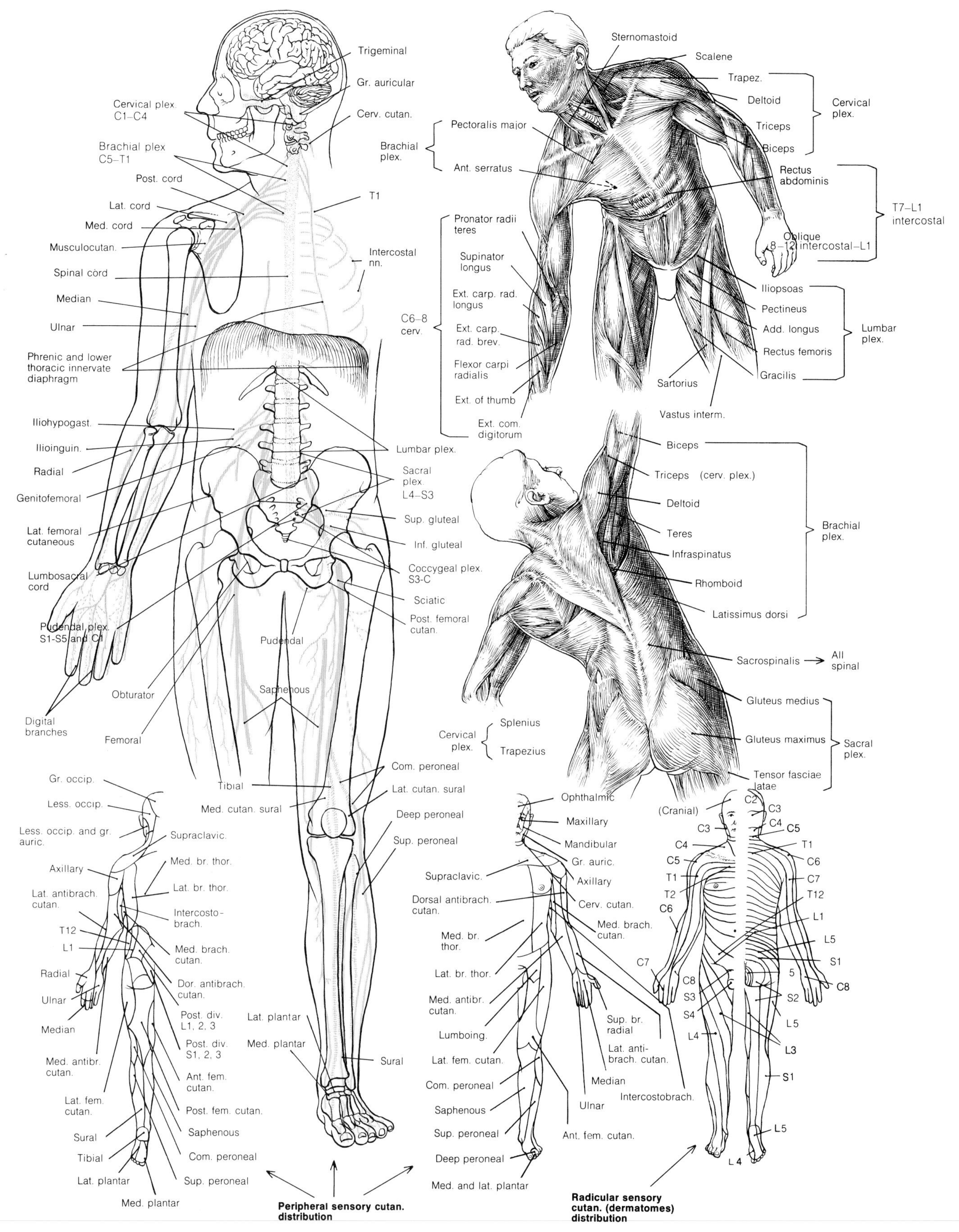

Plate 24.
Peripheral Nerves, Including Cervical, Brachial, and Lumbosacral Plexuses

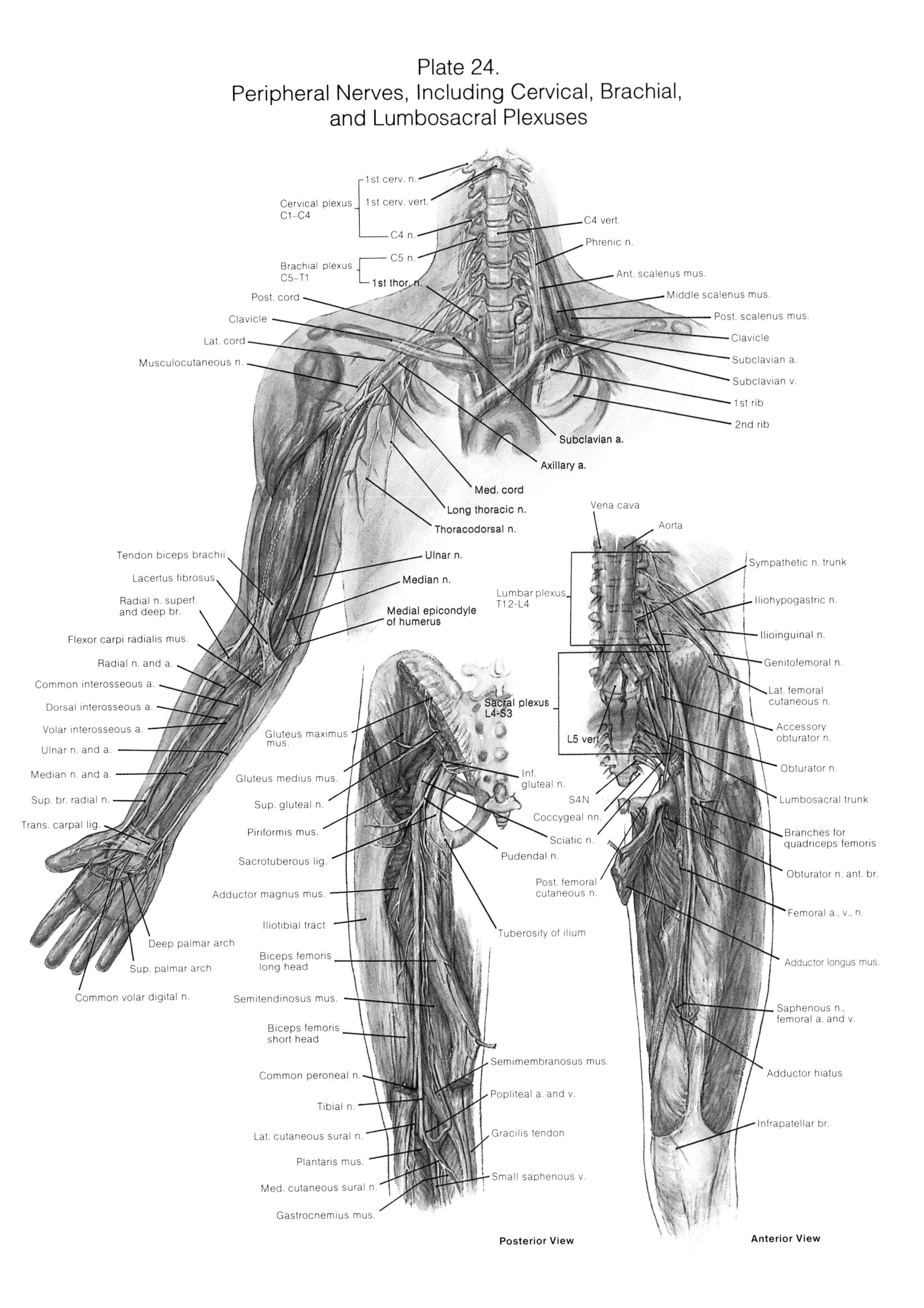

and lateral branches to the multifidae muscles; the lateral branches also supply the skin over the medial part of the gluteus maximus. The anterior primary divisions form part of the lumbosacral plexus. The white rami (parasympathetic) and the gray rami (sympathetic) form part of the sacral nerves, the first passing from the second, third, and fourth sacral nerves to the pelvic and lower abdominal viscera through the hypogastric plexus, and the second joining each sacral nerve from the sympathetic trunk.

The sacral plexus lies against the piriformis muscle on the posterior wall of the pelvis. Several structures like the colon and hypogastric muscles lie in front of it. The sacral plexus is formed by the fourth and fifth lumbar nerves, and the first, second, and third sacral nerves. The upper four posterior divisions (L4–L5–S1–S2) are fused to form the common peroneal nerve; all five of the anterior divisions (L4–L5–S1–S2–S3) participate in the formation of the tibial nerves. The posterior division of S3 with some branches of the anterior divisions of S2 and S3 participate in the formation of the pudendal plexus. The sacral plexus (L5–S5) forms the origin of (1) the superior gluteal nerve (to the gluteus maximus, medius, and minimus; tensor fasciae latae; and piriformis muscles), serving abduction and inward rotation of the leg, flexion and extension of the leg at the hip, and outward rotation of the leg; (2) the inferior gluteal and sciatic nerves (obturator internus, gemellar, quadratus femoris, biceps femoris, semitendinosus, and semimembranosus muscles), serving outward rotation and flexion of the leg at the hip; (3) the peroneal nerve with two branches, the deep innervating the tibialis anterior, extensor digitorum longus, and extensor hallucis brevis muscles, which dorsiflex and supinate the foot and extend the toes, and the superficial innervating the peroneus muscle (pronation of the foot); (4) the tibial nerve (to the gastrocnemius, soleus, posterior tibial muscles, and flexors of the toes), responsible for the plantar flexion of the foot; (5) the pudendal nerve, innervating the perineal muscles and the sphincters, being responsible for the closure of sphincters of the pelvic organs, contraction of the pelvic floor, and participation in the sexual act.

The sensory innervation of the skin is either related to a proximal, dermatomic (root) or to a distal, cutaneous (peripheral nerve) distribution as illustrated in Plate 23. By charting the sensory deficit, the site of lesion or disease can be localized. It is of considerable importance that there is an overlap of three to five segments to any given dermatome. To denervate completely the dermatomic area on the skin, two segments above and two segments below the given area have to be denervated to produce complete anesthesia in a given segment.

Different grades of sensory disturbances, from hypesthesia to complete anesthesia, involving either the cutaneous radicular or the peripheral nerve fields may be present. Irritative phenomena, like spontaneous pains or dysesthesias, may be present, indicating overstimulation of sensory modalities along root distribution or in the territory of a peripheral nerve. These pains and dysesthesias, when they originate from an irritable focus in a posterior spinal root, are projected through the corresponding peripheral nerves. The brain will perceive, as source of the incoming pain, the area of the dermatome that is supplied by the affected spinal root. The anatomical regions to which pains are projected when a spinal root is involved are well defined. For example, irritation of the second lumbar root will result in the pain being projected to the anterior aspect of the thigh; the third lumbar root is projected into the middle anterior aspect of the thigh and the inner knee; the fourth lumbar root to the lateral aspect of the thigh, front and medial side of the leg, and the great toe; the fifth lumbar root to the lateral side of the leg, dorsal and lateral sole of the foot, and so on.

It is of interest that the sciatic nerve is the largest in the body. In the thigh, one sheet surrounds the two separate nerves, the common peroneal and the tibial nerves. The course of the sciatic nerve after leaving the pelvis to the greater sciatic foramen, usually below the piriformis muscle, is between the greater trochanter of the femur and the tuberosities of the ischia, along the posterior surface of the thigh, and then to the popliteal space, where it terminates by dividing into the tibial and common peroneal nerves.

10 The Central Nervous System

Melvin H. Epstein, M.D.
Donlin M. Long, M.D.

THE BRAIN

The brain can be divided into three major parts: the brain stem, the cerebellum, and the cerebrum. The brain stem can be further subdivided into medulla, pons, and midbrain. The medulla is that part of the brain which joins the brain to the spinal cord. The pons and midbrain are successively more rostral.

Gross Anatomy

The cerebrum is the newest part of the brain and that part which is responsible for highest mental function. The cerebrum is subdivided into lobes, four in all on each side. Grossly, the brain is exactly the same on both right and left, while functionally there are differences between the right side of the brain and the left. Most anterior are the frontal lobes, next come the parietal lobes, and the most posterior lobes are the occipital. The temporal lobe is tucked like the thumb of a mitten into the area just above the ear. Each of these lobes has special functions.

The outermost surface of the brain is composed of nerve cells and is called the *cortex.* The processes of these nerve cells passing upward and downward comprise the white matter, which is the greatest volume of the brain. Deep inside the brain, just above the brain stem, are several masses of nerve cells with very important functions. The most important of these is the thalamus, which is a way station for relay of messages to the cortex and has important functions in processing information as well. Another of these buried areas of gray matter is the hypothalamus, which has important connections in behavior and in hormone function. The third of these areas of nerve cells is called the *basal ganglia.* These cells are very important in coordination of motor movement.

The cerebrum sits on the brain stem, and the cerebellum, especially important in coordination of motor movement, is located in the angle between the brain stem and cerebrum. The brain stem has many important functions. Most of the cranial nerves come from the brain stem, and all of the fiber tracts passing up and down from peripheral nerves and spinal cord to the higher parts of the brain must traverse the brain stem. The brain stem is especially important in control of subconscious and reflex activities such as breathing, heart rate, and blood pressure.

The Cerebrospinal Fluid

If the brain is cut in cross section, it is seen to have four cavities within it. Inside the cerebrum there are large lateral ventricles that connect in the midline to the third ventricle. The third ventricle is connected by a very narrow passage called the *aqueduct* to the fourth ventricle, which lies between the brain stem and cerebellum. Within these ventricles a structure called the *choroid plexus* is located. The choroid plexus produces the cerebrospinal fluid, which is a clear, watery fluid that both supports the brain and provides its extracellular fluid. This fluid circulates through the ventricles, leaves the fourth ventricle, and descends in the spinal canal to circulate around the spinal cord and the spinal nerves, returning upward to pass over the entire surface of the brain to be absorbed into the veins.

Functional Anatomy of the Brain

The frontal lobes are important in two major areas. The anterior portion of the frontal lobe is called the *prefrontal cortex* and is especially important in highest mental function and in the determination of personality. The posterior part of the frontal lobe controls motor movement and is divided into a premotor and motor area. In the motor cortex are located the nerve cells that actually produce movement, and in the premotor area are several portions of the brain that modify movement. The frontal lobe is divided from the parietal lobe by the central sulcus. Immediately behind the motor cortex is the primary sensory cortex. This controls sensation such as touch, pressure, and localization of objects that touch the skin. Just behind this primary sensory area is a very large association area that controls such fine sensation as judgment of texture, weight, size, and shape.

The occipital lobe is concerned with vision. On the medial surface of the occipital lobe is located the calcarine cortex, which contains the cells that have to do with primary visual reception. The remainder of the occipital lobes are association areas that help in the recognition of size and shape and color.

The temporal lobe has numerous important functions. The auditory cortex is located on the superior internal portion of the temporal lobe, and an area called the *hippocampus* forms the lobe's medial portion. This hippocampus and related structures are very important in behavior. The medial part of the temporal lobe is connected to the hypothalamus and then to the frontal lobes, with interconnections to many other parts of the brain.

On the left side of the brain in right-handed individuals is located the speech area. The hemisphere containing speech is called the *dominant hemisphere.* Motor speech is located at the base of the frontal lobe, in Broca's area. This is the part of the brain that controls the movement necessary for speech. On the lateral portion of the temporal lobe there is an important area that has to do with hearing speech, and in the base of the parietal lobe are association areas that have to do with understanding and carrying out the complex actions required for speech.

The portion of the brain most responsible for behavior is called the *limbic lobe.* The limbic lobe is not an anatomical lobe of the brain but a functional subdivision.

Blood Supply of the Brain

The brain gets its primary blood supply from the two internal carotid arteries and the two vertebral arteries. The vascular connections at the base of the brain form the circle of Willis, which is composed of a vascular loop consisting of the two anterior cerebral arteries and the anterior communicating artery in front and the posterior communicating artery and the posterior cerebral arteries behind. There are numerous variations and congenital abnormalities associated with the circle of Willis.

The internal carotid artery divides intracranially into two main branches. One is the anterior cerebral artery, which passes forward and immediately above the optic chiasm to enter the longitudinal fissue of the cerebrum. It furnishes blood to the major part of the medial aspect of the cerebral hemisphere and gives off several vital perforating branches at the base of the brain that supply blood to the head of the caudate, the anterior part of the lentiform nucleus, the internal capsule, the anterior columns of the fornix, and the anterior commissures. The loss of these important perforators leads to deep coma.

The middle cerebral artery is the larger of the two terminal branches of the internal carotid artery and passes laterally through the Sylvian fissure to the surface of the insula, where it divides into numerous parietal and temporal cortical branches. During its course through the Sylvian fissure it gives off important perforating arteries called *medial* and *lateral striate arteries,* which pass upward through the putamen of the lentiform nucleus, and it also supplies blood to the globus pallidus and the internal capsule. These arteries that frequently rupture in cases of spontaneous cerebral hemorrhage are known as the *arteries of Charcot.*

The vertebral arteries enter the intracranial cavity and traverse the base of the medulla, giving off two posteroinferior cerebellar arteries, which supply blood to the brain stem and the posteroinferior surface of the cerebellum. The vertebral arteries then join to become the basilar artery. The important perforating branches from the basilar artery to the remainder of the brain stem are vital to many life functions. The anteroinferior cerebellar artery arises from the basilar artery at the pontomedullary junction and gives blood to the anteroinferior surfaces of the cerebellum. There is frequently an important loop that passes into the internal auditory canal from the anteroinferior cerebellar artery and returns to the brain stem to supply blood to the pons. Close to the basilar summit arises the superior cerebellar artery, which has perforating branches to the brain stem and to the superior surface of the cerebellum.

The posterior cerebral arteries are the terminal branches of the basilar system. They also have extremely important perforating central arteries, which supply blood to the cerebral peduncle, the posterior perforated substance, the posterior part of the thalamus, and the mammillary bodies during their course around the brain stem. The posterior choroidal branches pass through the upper part of the choroid fissure, then to the posterior part of the tela choroidea of the third ventricle, and then to the choroid plexus. The posterior cerebral supplies blood to the uncus, hippocampal gyrus, medial temporal lobe, occipital pole, and a small portion of the posterior parietal lobe.

The Cranial Nerves

The olfactory nerve, or first cranial nerve, is the pathway taken by olfactory impulses from the nasal mucosa to the brain. The olfactory tract connects the olfactory bulb with the olfactory tubercle, where it divides into a

Plate 25.
The Five Senses of Consciousness

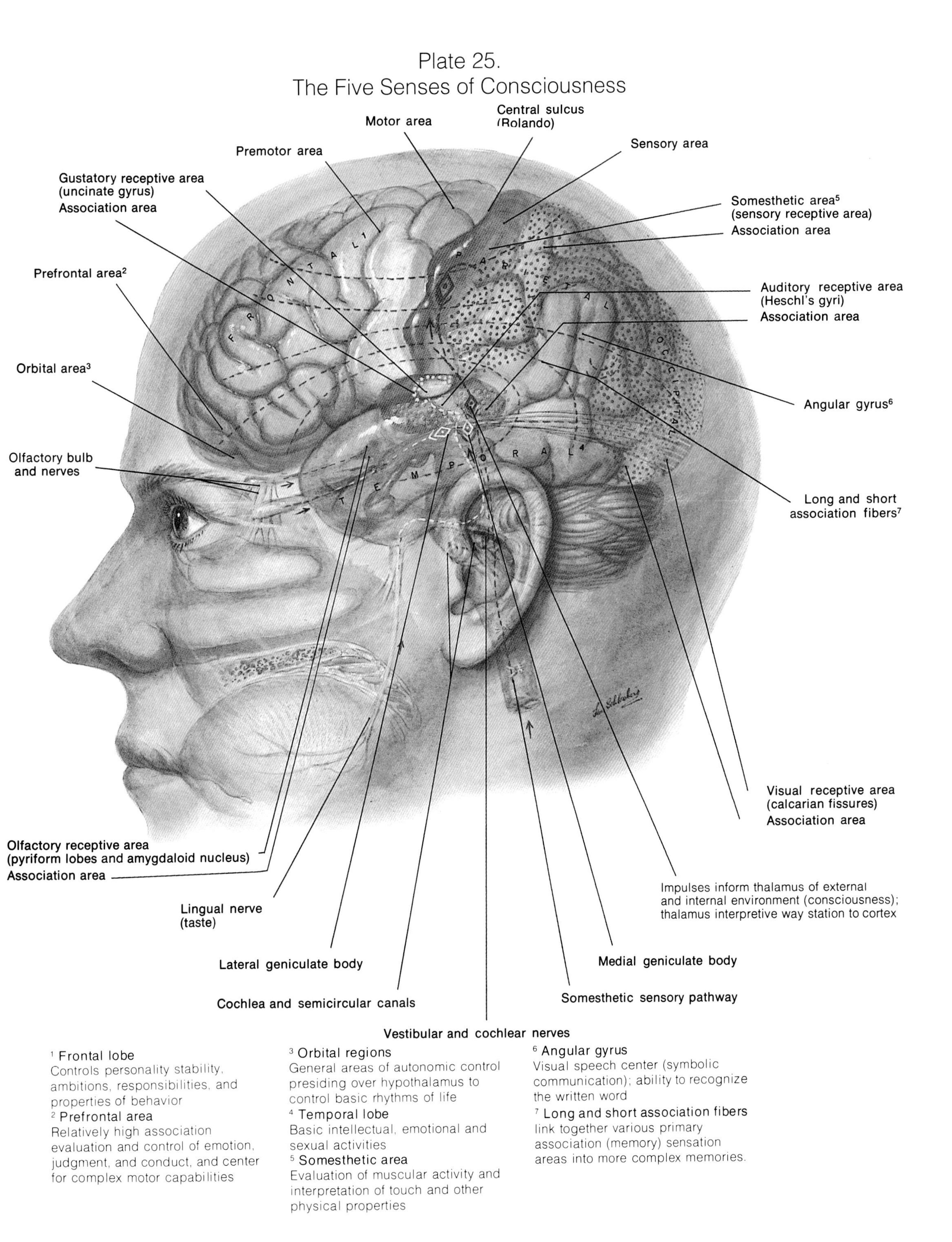

[1] Frontal lobe
Controls personality stability, ambitions, responsibilities, and properties of behavior

[2] Prefrontal area
Relatively high association evaluation and control of emotion, judgment, and conduct, and center for complex motor capabilities

[3] Orbital regions
General areas of autonomic control presiding over hypothalamus to control basic rhythms of life

[4] Temporal lobe
Basic intellectual, emotional and sexual activities

[5] Somesthetic area
Evaluation of muscular activity and interpretation of touch and other physical properties

[6] Angular gyrus
Visual speech center (symbolic communication); ability to recognize the written word

[7] Long and short association fibers
link together various primary association (memory) sensation areas into more complex memories.

Plate 26.
Vascular Supply of the Brain

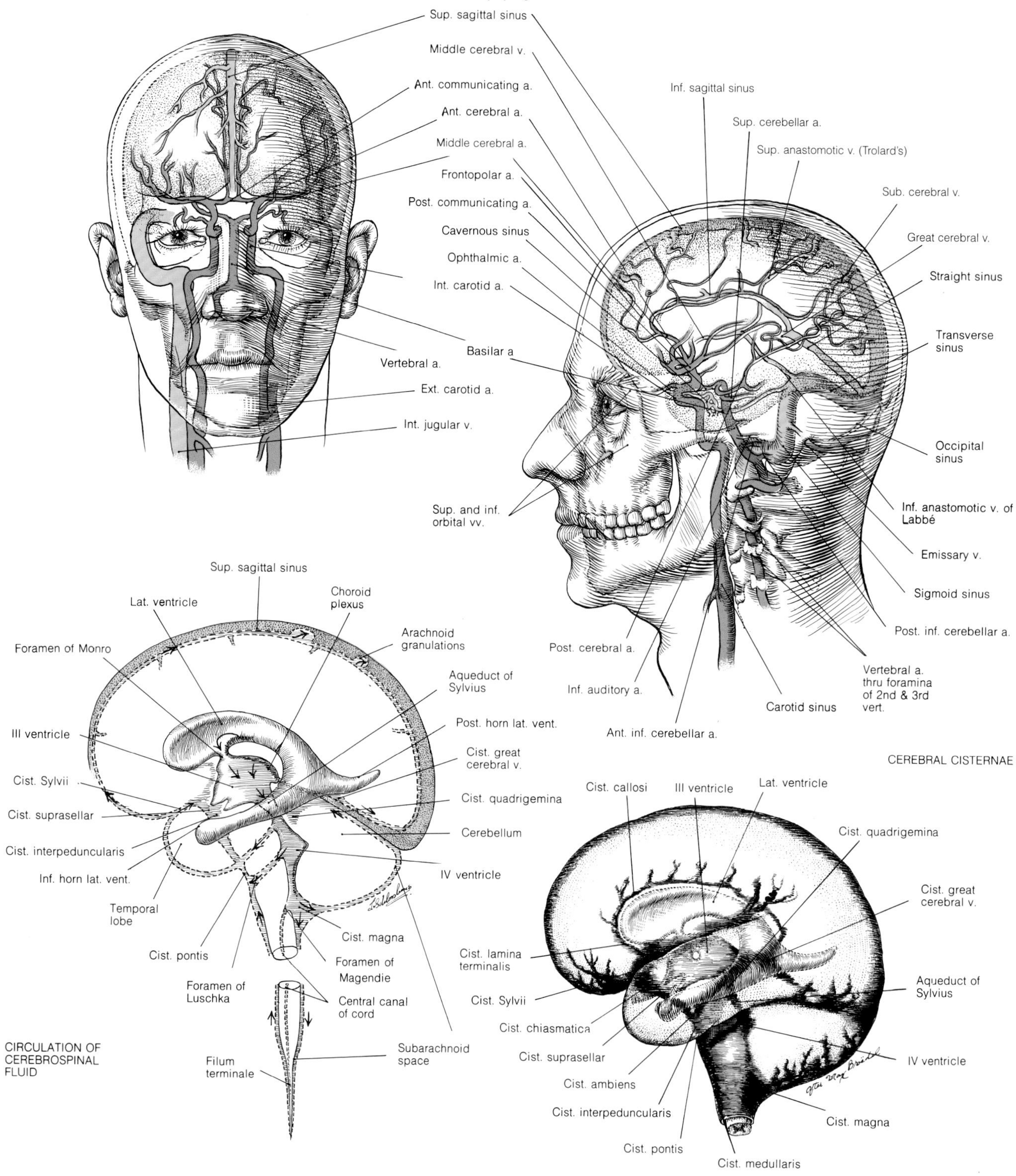

medial and lateral olfactory tract. The optic nerve, or second cranial nerve, lies just posterior and inferior to the medial olfactory tract. It carries information from the eye for vision and ocular reflexes. The third cranial nerve, or oculomotor nerve, arises at the ventral aspect of the mesencephalon and traverses through the cavernous sinus to the orbit. It supplies all the intrinsic ocular muscles and all extrinsic ocular muscles except for the lateral rectus and superior oblique. The parasympathetic fibers from this nerve innervate the ciliary muscle of the lens and the sphincter muscle of the pupil. The fourth, or trochlear, nerve supplies only the superior oblique muscle of the eye, and it arises just below the inferior quadrigeminal bodies of the brain stem. It emerges from the posterior aspect of the brain stem and passes around the lateral side of the cerebral peduncle into the margin of the tentorium and into the cavernous sinus, where it goes to the orbit. The fifth cranial nerve, or trigeminal nerve, is the largest cranial nerve, and it carries fibers that give sensation to the face and motor fibers to the muscles of mastication. It exits from the brain stem through the anterolateral surface of the pons.

The sixth, or abducent, nerve supplies the lateral rectus muscle of the eyeball and issues from the brain at the inferior border of the pons, just above the pyramid of the medulla. The seventh, or facial, nerve consists of two parts: the motor root, which supplies the superficial muscles of the scalp, face, and neck; and a smaller sensory root, which contains the afferent taste fibers for the anterior two thirds of the tongue and the afferent parasympathetic fibers for supply of the lacrimal and salivary glands. The facial nerve arises from the lateral aspect of the pontomedullary junction. The auditory nerve, or eighth nerve, is entirely sensory and consists of vestibular and cochlear divisions. The glossopharyngeal, or ninth, nerve is a mixed nerve consisting of an afferent part, which supplies the pharynx and tongue and the carotid sinus and body, and the efferent part, which supplies the stylopharyngeus muscle. It arises from the medulla by five or six fine rootlets which are attached to the side of the medulla oblongata, close to the facial nerve.

The vagus, or tenth, nerve is also a mixed nerve, which contains a large number of parasympathetic fibers and passes through the neck and thorax into the abdomen. It supplies afferent fibers chiefly to the pharynx, esophagus, stomach, larynx, trachea, and lungs. It is attached by numerous rootlets to the side of the medulla, in series with the glossopharyngeal nerve above and the accessory nerve below. The rootlets unite to form a single tract which exits from the cranial cavity through the jugular foramen. The accessory nerve, or eleventh cranial nerve, consists of bulbar and spinal portions. It arises in series with the vagus and glossopharyngeal nerve and controls motor function of the sternomastoid and the trapezius muscles. The twelfth, or hypoglossal, nerve is a predominantly efferent nerve that supplies all the muscles of the tongue, both intrinsic and extrinsic, except the palatoglossus muscle. It arises from numerous rootlets from the anterior portion of the medulla oblongata. The rootlets arrange themselves in double bundles and unite in the anterior condylar canal, where they emerge from the cranial cavity.

Circulation of Cerebral Spinal Fluid

Most of the cerebral spinal fluid is formed within the lateral ventricles of the brain by the choroid plexus. Cerebral spinal fluid is a clear, colorless liquid of low specific gravity that in health has between two and three lymphocytes per cubic millimeter. Its total volume is between 100 and 140 ml in adults, and the normal pressure varies from between 70 and 200 mm of water with the patient on his side. Total protein in the adult varies from 20 to 45 mg percent, with glucose varying from 50 to 75 mg percent. Chlorides are between 120 and 230 m eq/lit.

From the lateral ventricles, fluid traverses the interventricular foramina into the third ventricle. Here, presumably, the choroid plexus of the third ventricle contributes fluid, which then passes through the aqueduct of Sylvius into the fourth ventricle, where further additions are made by the choroid plexus in the roof of the fourth ventricle. The fluid then escapes into the subarachnoid space through a median aperture called the *foramen of Magendie* and lateral apertures called *foramina of Luschka.* Some fluid passes downward into the spinal subarachnoid space, but the major portion rises through the tentorial notch and finds its way slowly over the surface of the hemispheres to be absorbed mainly through the arachnoid villi and granulations into the venous system. There appear to be other mechanisms of CSF absorption, mainly through the perineural lymphatics, and, in addition, in abnormal pressure states it appears possible for CSF to be absorbed through the ependyma. Each blood vessel, as it enters the brain tissue, incorporates a prolongation of the subarachnoid space by which cerebrospinal fluid can come in contact with the neurons themselves. Any obstruction to the ventricular system, either by blockage of the foramen of Monro, the aqueduct of Sylvius, or of the foramina of Magendie and Luschka, will create a non-communicating type of hydrocephalus in which the ventricular system will dilate. The pressure within the ventricular system in hydrocephalus can increase in an acute and fatal manner. If absorption is interfered with, there develops a hydrocephalus that is classified as communicating.

THE SPINAL CORD

The spinal cord is a continuation of the lower part of the brain stem, which descends in the bony vertebral col-

Plate 27.
Base of the Brain and of the Skull

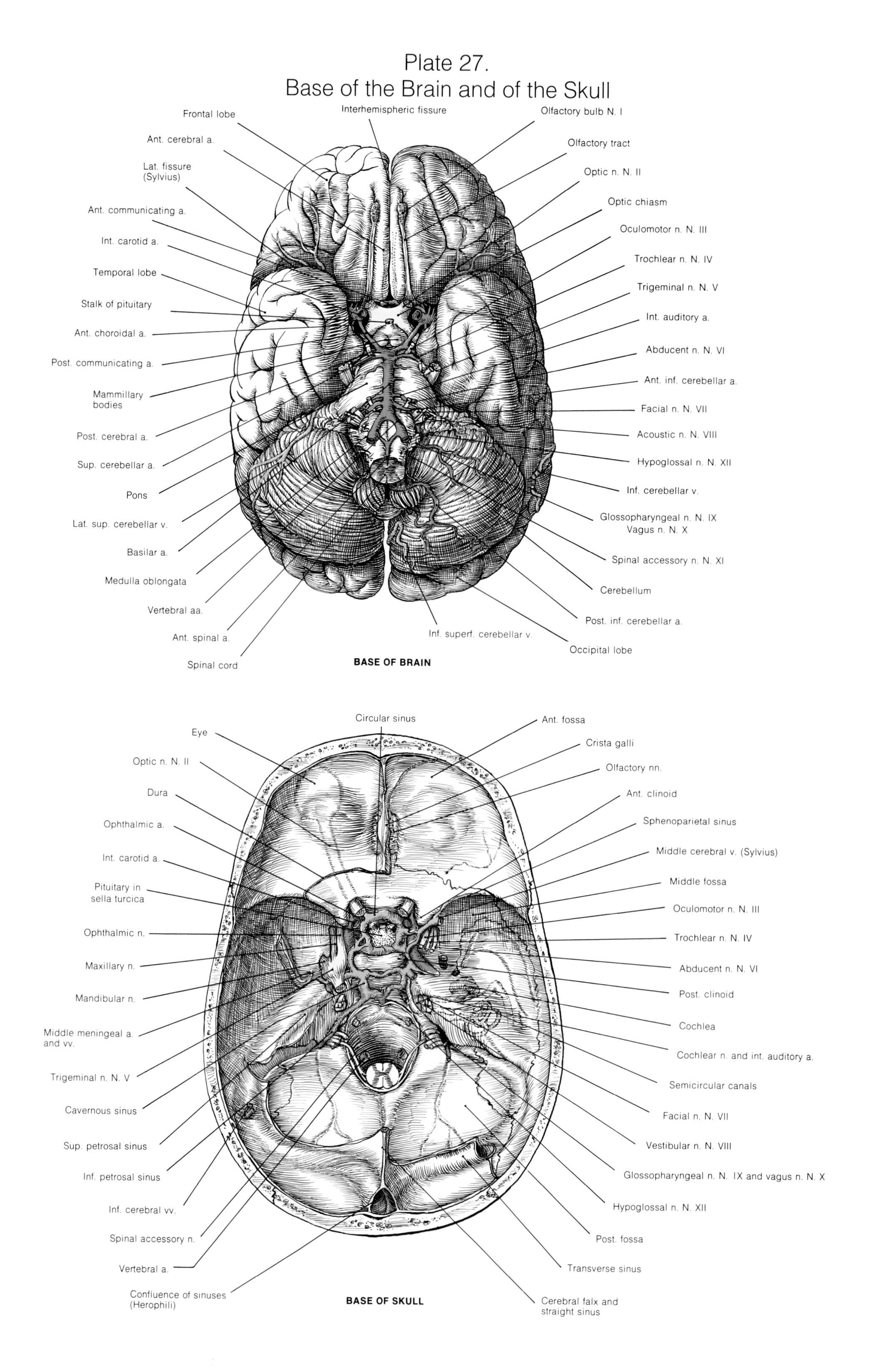

umn giving off thirty-one pairs of nerves. The spinal cord ends between the first and second lumbar vertebral bodies as the conus medullaris. Below this level the nerve roots are called the *cauda equina.*

The spinal cord is divided into anterior and posterior halves. This division is recognized by the dentate ligaments, which attach the lateral sides of the spinal cord to the surrounding dura. The posterior half of the spinal cord is further divided by the posterior median longitudinal fissure. The posterior spinal roots enter the spinal cord at approximately the midpoint of this posterior quadrant. The anterior half of the spinal cord is also divided by the anterior longitudinal fissure, which runs down the midline. In this fissure is located the anterior spinal artery, which is the major blood supply to the spinal cord. The anterior spinal roots arise from the anterior quadrant and pass laterally to join the posterior spinal roots at the dura mater, to continue outward as peripheral nerves. Each of these small roots is made up of several rootlets, which join together to form the roots. There are eight cervical nerves on each side, twelve thoracic, five lumbar, and five sacral. The small coccygeal that makes the thirty-first pair is inconstant. A long fibrous structure called the *filum terminale* attaches the conus medullaris to the sacrum. In the cervical region and at the lower end of the spinal cord, the spinal cord swells as the cervical and lumbar enlargements. These are the areas where the large nerves that make up the brachial plexus and the lumbosacral plexus arise. These plexuses provide the supply of sensation and movement to the arms and legs.

The blood supply of the spinal cord comes primarily from the anterior spinal artery, which arises from the vertebral artery and then descends in the anterior median sulcus. At many levels in the spinal cord, small blood vessels enter with the nerve roots and anastomose with the anterior spinal artery.

Functional Anatomy of the Spinal Cord

The posterior medial portions of the spinal cord are called the *posterior columns.* These transmit such sensations as position, joint sense, and pressure. Lateral to the posterior columns, but still behind the dentate ligament, are located the corticospinal tracts. These are the motor tracts that control movement. The outer portion of the spinal cord is white matter, and the nerve cells or gray matter are located internally, in structures called *horns.* In front of the dentate ligament in the anterior quadrants are located the anterior horns. The large nerve cells of the anterior horns are the final common path for all motor activity and supply the impulses that cause movement in all muscles of the body except those supplied by the cranial nerves. The nerve cells that have to do with sensation are in the posterior portion of the spinal cord and are called the *posterior horns.*

The spinal cord is surrounded by a tough fibrous structure called the *dura mater.* This covers the spinal cord and all of the spinal nerves. The spinal nerves exit from the spinal cord, pierce the dura, and descend to reach the nearest neural foramen. The foramen is the opening between the vertebral bodies through which the nerves exit. In this foramen is located a small structure called the *dorsal root ganglion,* in which the cells that supply sensation to the body are located. As soon as the nerve leaves the bony protection of the neural foramen, it is termed a *peripheral nerve.* The spinal nerves are numbered according to the vertebral body with which they are associated. In the cervical region, all of the nerves leave above the vertebral body of the same number except the eighth cervical nerve, which exits below the body of the seventh vertebra. Below that level all of the nerves exit below the vertebral body of the same number.

It is possible to locate the area of the spinal cord by the relations of the spinal segments with the vertebra. Between the level of the second cervical and the tenth thoracic spinous processes, adding two to the number of the spinous process felt will give the underlying spinal cord segment. The eleventh and twelfth spinous processes overlie the five lumbar segments and the first lumbar spinous process overlies the five sacral segments.

Plate 28.
Cerebrospinal Axis and Cervical Cord

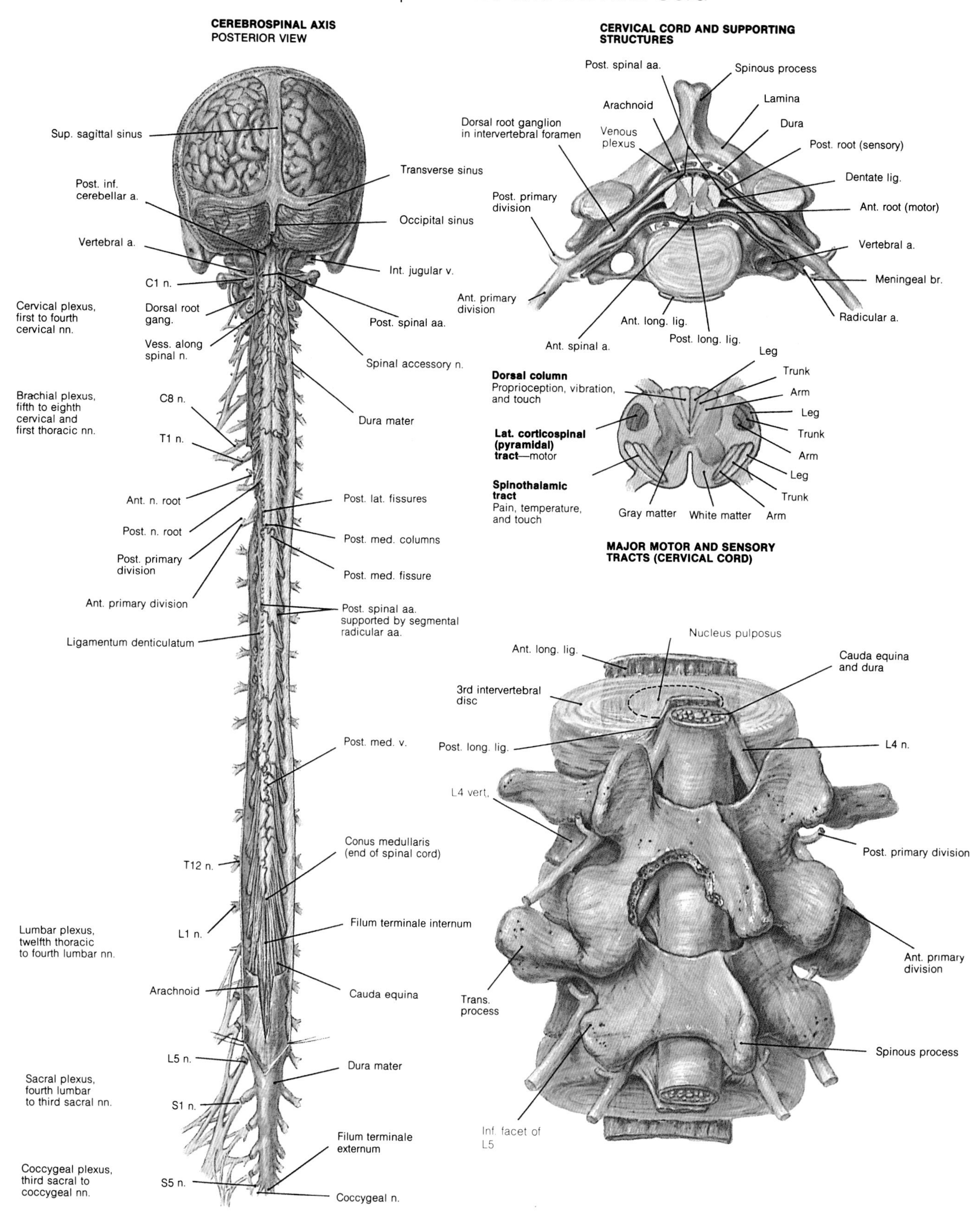

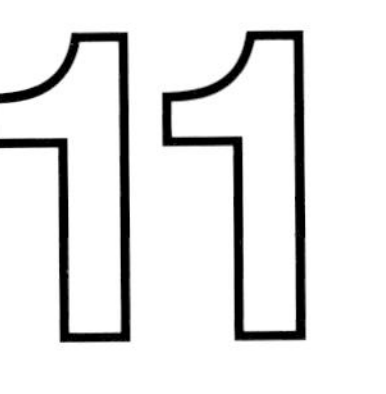

The Lymphatic System

James P. Isaacs, M.D.

The lymphatic system is an active component of the circulatory system that drains peripheral tissues in parallel with venous return.

Virtually all structures in the body possess a fine reticulum of blind-ended lymphatic channels which are lined by endothelial cells. These cells pass many different chemical and biological substances from tissues in the formation of lymph.

Lymph is an isotonic, relatively fat-rich, protein-poor fluid that is carried from its reticular origin through afferent tubular lymphatics to lymph nodes.

Each lymph node has an arterial blood supply and venous drainage, with lymphoid germinal centers that add lymphoid cells to circulating lymph and blood. Sinuses of each lymph node collect into an efferent tubule, which joins with other tubules to form larger lymphatic trunks. The trunks empty into two main lymphatic ducts, the thoracic duct and the right lymphatic duct.

The lymphatic system originates embryonically from transformed venous endothelium as a series of paired and unpaired sacs. The jugular and sciatic lymphatic sacs are paired, arising respectively from jugular and iliac veins; the retroperitoneal sac, unpaired, arises from the inferior vena cava and mesonephric veins. The unpaired cisterna chyli, which originates from the Wolffian ducts, becomes the drainage center for retroperitoneal and sciatic lymphatic sacs, jejunal and ileal fatty lymphatics or lacteals, and bilateral descending intercostal lymphatic trunks. The cisterna chyli occupies a strategic retroperitoneal position in the right upper lumbar paravertebral gutter. The thoracic duct empties the cisterna chyli, passing through the diaphragm into the right chest, crossing into the left chest at the fifth thoracic vertebra, and extending to the lower left neck. Here it drains into the junction of the left subclavian and internal jugular veins. Near this venous junction on the left or right side of the neck, three lymphatic trunks (subclavian, inferior jugular, and bronchomediastinal) join the thoracic duct or right lymphatic duct respectively.

Musculature is present in the walls of lymphatic trunks and ducts in increasing amounts as these vessels get larger. Contraction of this intrinsic muscle in the presence of one-way valves exerts a pumping action on the lymph. Contraction of surrounding muscle masses also assists in lymph pumping. Lymph flow rate from the thoracic duct terminus is 50 to 100 cc per hour at a rest pressure near zero. The thoracic duct pressure and flow rise sensitively with exercise, increased body temperature, and alimentation.

The lymphatic system is also a defense system with limited capacity. When pathologically overwhelmed, it is a route for spread of bacterial infections, parasitic infestations, tumor metastases, foreign body contaminations, inflammatory degenerations, and chemical absorptions. The lymphatics can be obstructed by accumulation of particulate materials, by lymphangitis, and by lymph clotting, thrombosis, and embolization. The lymphatic valves are injured in the same manner that venous valves are injured by phlebitis and venous thrombosis. Traumatic division of larger lymphatic vessels also occurs. The various obstructions and divisions create troublesome or sometimes fatal syndromes, which include chyle thorax, chyle peritoneum, and lymphedema of extremities, genitourinary organs, and so forth.

The lymphatic system has other forms of pathology: neoplasia (Hodgkin's disease, lymphomas, lymphosarcoma, cystic hygroma, etc.) and metaplasia and hyperplasia (pyogenic, granulomatous, and other infections). Some lymphatic vessels may be congenitally absent (congenital lymphedema).

Plate 29. Lymphatics—Head, Neck, and Chest

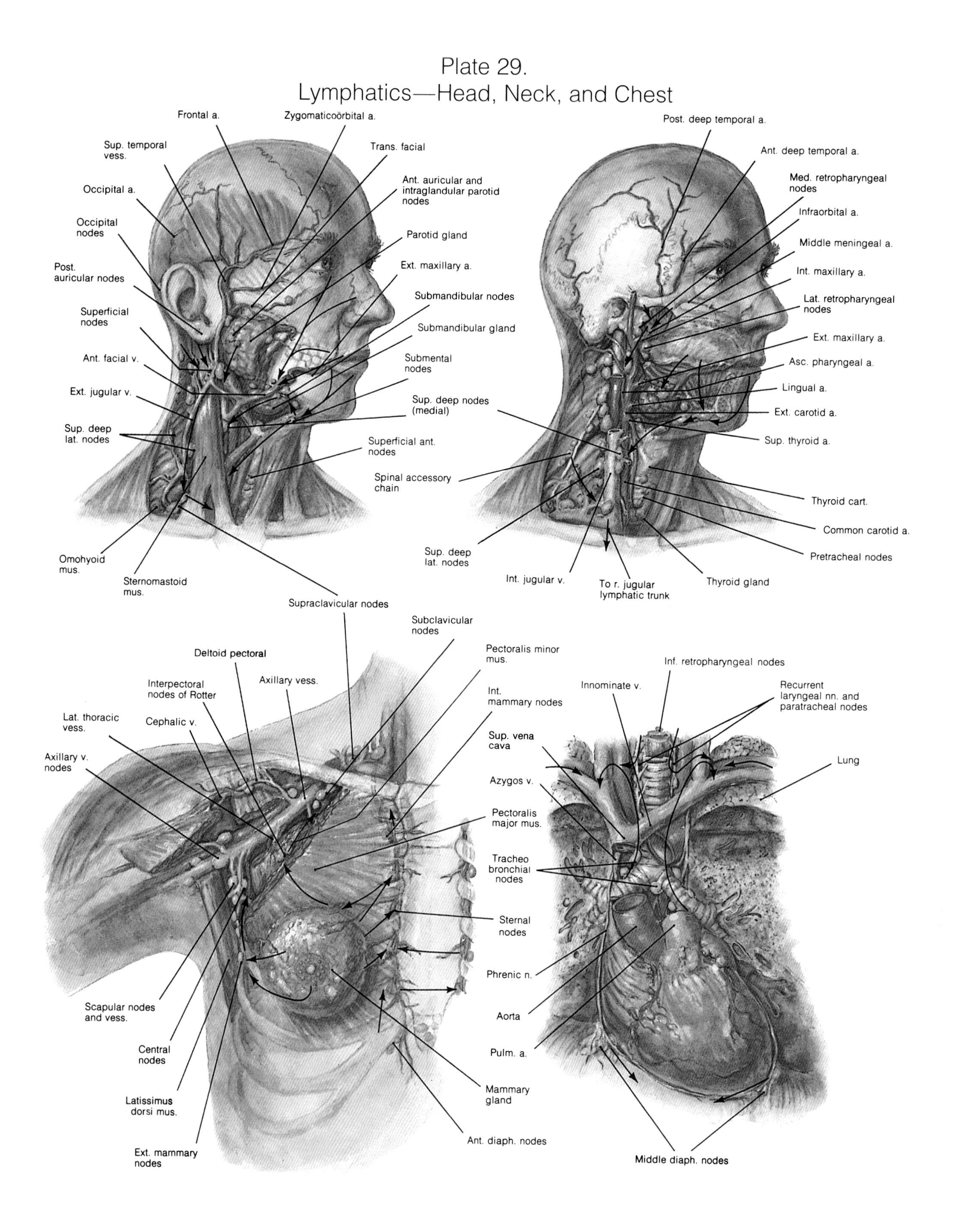

Plate 30. Lymphatics—Esophagus and Stomach; Colon, Rectum, and Anus

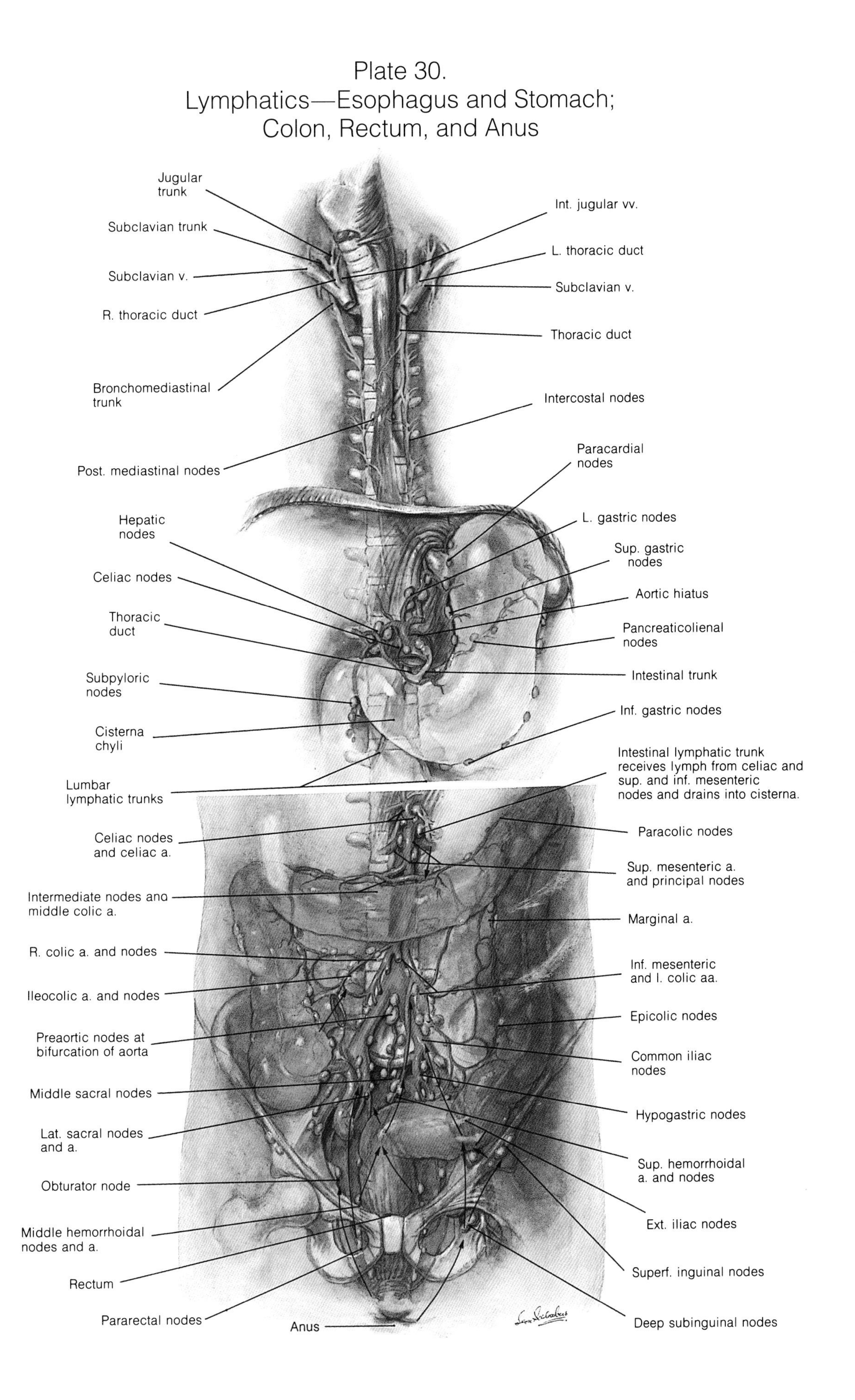

Plate 31. Lymphatics—Genitourinary System; Liver and Bile Ducts

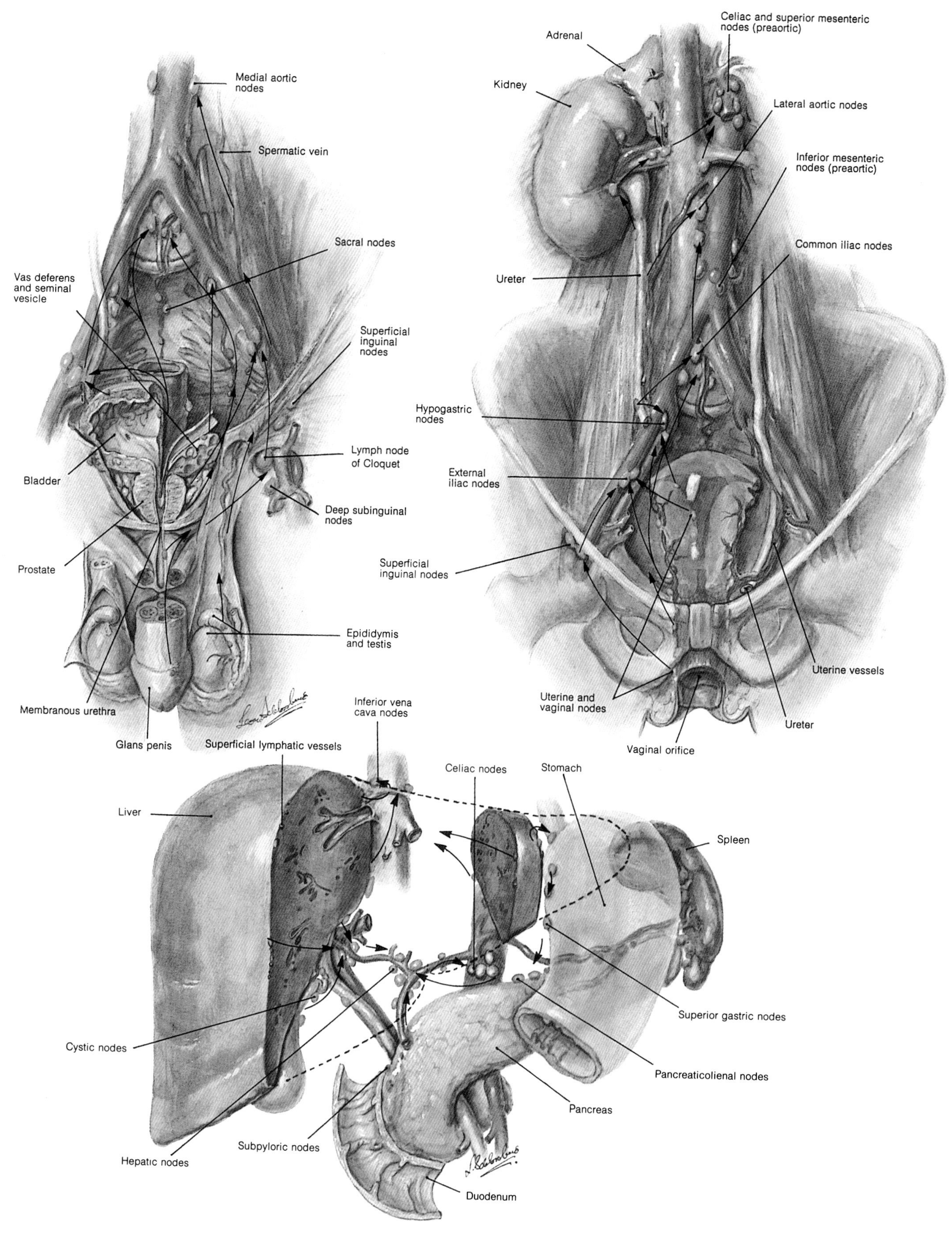

12 The Eye and the Mechanism of Vision

Charles E. Iliff, M.D.

THE LIDS, EXTRAOCULAR MUSCLES, AND LACRIMAL APPARATUS

The eye is surrounded by the bones of the skull on all sides except the front, where it is protected by the lids. The lids contain a platelike fibrous tissue—the tarsus—which gives inner support and added substance to this partially muscular curtain.

The lids are fastened nasally and laterally to the bony orbital wall by fibrous ligaments, so that when the orbicularis muscle contracts, the lid fissure is closed. This motion is voluntary in the wink and is reflex in the blink; these mechanisms are important in helping protect the globe from injury and for the lubrication of the cornea by spreading of tears, mucus, and other secretions from the glands of the conjunctiva. The lids are opened by the levator palpebrae muscle, which is the primary retractor of the upper lid, and by the secondary elevators of the lid: Müller's muscle, the superior rectus, and the frontalis muscle.

The eye rides in its bony socket on a cushion of fat through which run tendons, muscles, nerves, and blood vessels. The fat gently supports the globe, and the ligaments limit its motion.

Of the six extraocular muscles, the four recti have an origin around the optic foramen and insert on the sclera of the anterior portion of the globe. There are two oblique muscles. The superior oblique has its origin (with the four rectus muscles) at the annulus of Zinn, passes forward to the trochlea (pulley) on the anterior superonasal wall of the orbit, and from here passes beneath the superior rectus muscle and inserts into the sclera, temporal to and above the posterior pole of the globe. The inferior oblique muscle originates from the anterior orbital wall near the lacrimal fossa and passes backward to insert on the globe below the horizontal meridian and slightly temporal to the posterior pole. These six muscles working together give each globe its full range of motion, which permits the two eyes to work together and allows the visual axes to rotate as far as the limits provided by the orbital rims.

The lacrimal gland in the superotemporal quadrant of the orbit secretes tears (lacrimae), which lubricate the conjunctiva and cornea. The tears pass from the lacrimal ducts in the upper temporal conjunctiva and then across the globe to exit through the lacrimal puncta, the single small openings on the nasal edge of the upper and lower lid. The tears are normally kept from running over the lid margins by the fatty secretion from the Meibomian glands and glands of Moll, and are somewhat reduced in amount by evaporation as they pass to the lid puncta. Here they are funneled along the upper and lower lacrimal ducts into the lacrimal sacs, and from there through a bony canal to the mucous membrane of the nose, where they are evaporated by the air passing over the turbinates during respiration.

VASCULAR SUPPLY, LYMPHATICS, AND MAJOR NERVES

The ophthalmic artery, a branch of the internal carotid artery, supplies most of the orbital structures, including the eye and its internal structures. There are a few branches of the internal maxillary artery that supply the inferior rectus and oblique muscles, the lacrimal sac and gland, and lower lid. The venous system mainly drains backward into the cavernous sinus, but there are also many anastomoses with the facial veins and veins of the nasal cavity.

The lids have lymphatic drainage channels, but there are no lymphatics in the globe, and they are not normally found in the orbit. However, lymphangiomatous tumors do occur in the orbit, supposedly beginning in the lids and extending backward.

The nerves of the eye and orbit include the second cranial nerve—the optic nerve—which is actually an extension of the brain. The third cranial nerve—the oculomotor nerve—supplies the motor impulses for the levator muscles of the lid; the superior, medial, and inferior recti; and the inferior oblique muscles of the globe. The third nerve also sends off a motor root to the ciliary ganglion, which furnishes the autonomic innervation to the muscles within the globe, including the constrictor muscle of the iris.

The fourth and sixth cranial nerves are also motor nerves, and they supply the superior oblique and external rectus muscles respectively.

The fifth cranial nerve—the trigeminal—provides the sensory mechanism for the orbital structures, and in addition it supplies the surrounding area of the face and sinuses.

The seventh cranial nerve innervates the orbicularis muscles of the lids. The autonomic system controls the sphincter and dilator muscles of the pupil, the ciliary body, the operation of the lacrimal gland, and the smooth muscles of the lids and orbit.

ANATOMY OF THE EYE (THE GLOBE)

The globe itself measures about 24 mm in diameter. The anterior portion of the globe consists of the cornea, the curved transparent segment that is about 11 mm in diameter. The cornea is composed of avascular parallel layers of fibrous tissue, covered on the outside with epithelial and on the inside with endothelial cells. The posterior portion of the globe is the opaque sclera, which consists of a tough fibrous coat made up of connective tissue, elastic fibers, and blood vessels.

Behind the cornea is the anterior chamber, which is filled with a clear fluid, the aqueous humor. The posterior boundary of the anterior chamber consists of the crystalline lens and the iris. The lens is clear, and its shape is changed by the action of the ciliary muscle and by lens elasticity to enable the eye to focus on distant and near objects. The opening in the iris, the pupil, controls the amount of light entering the eye by the action of the constricting and dilating muscles of the iris. The lens is supported in the eye by fine fibers—the zonular fibers—which suspend the lens from the ciliary body. (These fibers, encircling the lens equator, are collectively called the *zonules.*) The ciliary body, in addition to being the base for the zonular fibers, secretes (or excretes) the aqueous fluid, which passes forward through the pupillary space into the anterior chamber and then exits through the trabecular meshwork and Schlemm's canal, which are located in the angle of the anterior chamber. The angle is formed by the attachment of the peripheral portion of the iris to the supporting sclera.

Just beneath the sclera, and lining the posterior portion of the globe, is the choroid, which carries the main vascular supply to the outer layers of the retina. The retina, which is a highly differentiated structure, with its receptor elements—the rods and cones—connects with the brain along the visual pathway of the optic nerve. The retina is a complex structure composed of an outer layer of pigment epithelium and an inner sensory epithelium. The sensory retina includes the rods and cones—the visual cells—which are supported by a connective tissue framework that carries ganglion cells, connecting cells, and nerve fibers; these fibers join to form the optic nerve, which carries the visual images to the brain. When the retina is observed with an ophthalmoscope, the optic nerve head (the optic disc) is seen as an oval structure with a pinkish hue, but of much lighter color than the surrounding retina. The ophthalmic artery and vein emerge from the nerve head to course over the retina and supply its superficial layers. The macula—a small central area containing a pit (the fovea), located near the posterior pole of the globe—is an area of the retina in which cones predominate and the rods are few; this is the area of most acute vision.

Within the cavity of the globe is the vitreous, which is a transparent, semigelatinous substance that provides an inner supporting structure central to the retina, choroid, ciliary body, and lens.

The optic nerve from each eye runs backward in the orbit and through the optic canal and then joins with the other optic nerve at the optic chiasm, where the temporal half of the fibers (from the lateral portion of retina) pass backward to the lateral geniculate body and from there to the visual cortex. The fibers from the nasal retina of each eye pass back to the optic chiasm and cross over (decussate) to the opposite optic tract (in an *X*-like manner) to join with the temporal fibers from the other eye, and with these fibers pass backward to the geniculate body and then to the visual cortex.

Plate 32.
The Eye and the Orbit

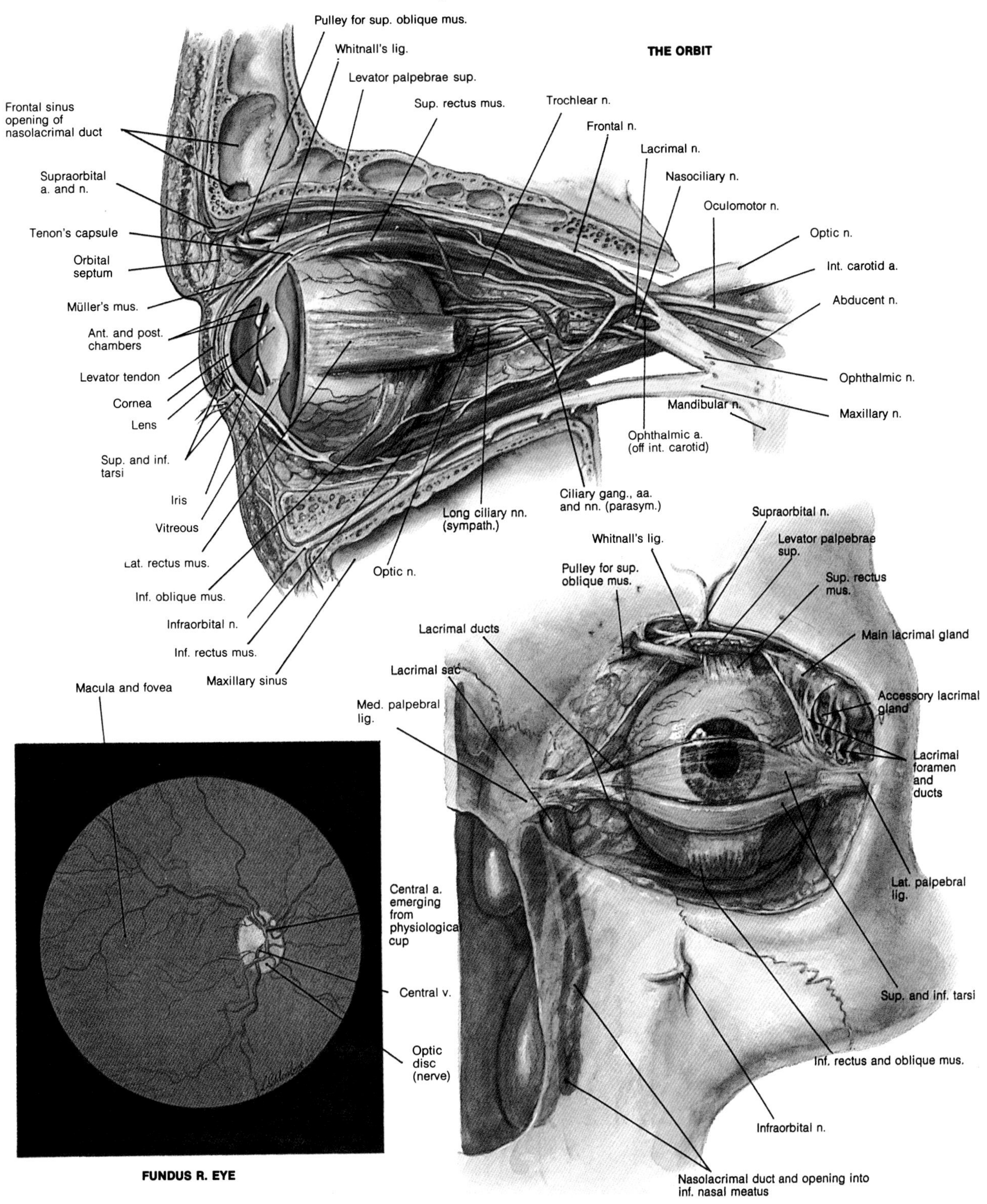

Plate 33.
The Eye—Vision

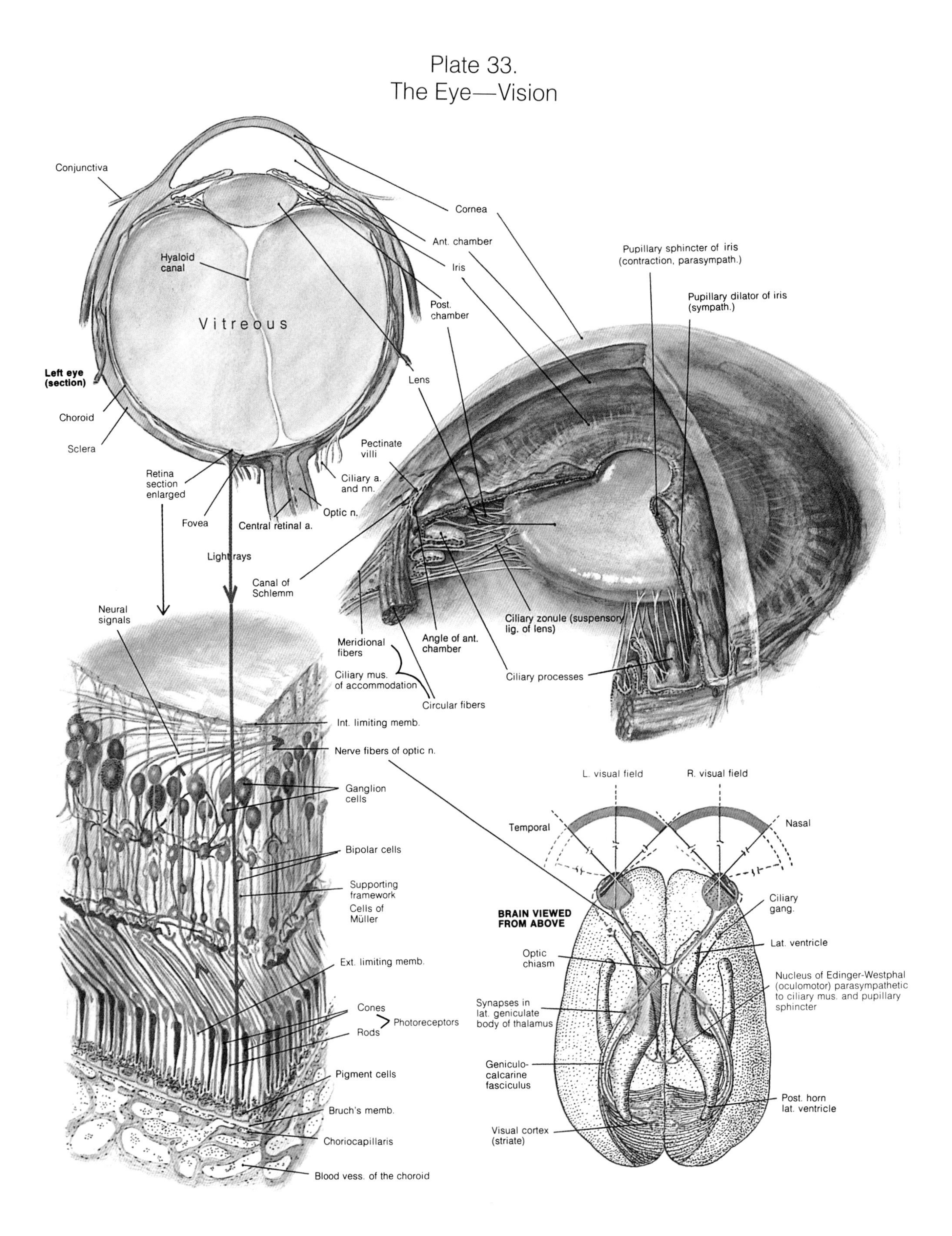

13 The Ear

George T. Nager, M.D.

The ear may be divided into the external, middle, and inner portions. The external ear consists of the auricle, or pinna, and the external auditory canal. The auricle, which, in man, does little to increase the sensitivity of hearing, is made of a cartilaginous framework covered by skin, and is attached to the underlying temporal bone. Its sensory nerve supply is from the fifth, seventh, and tenth cranial nerves and from the cervical plexus.

The external auditory canal is bayonet-shaped and measures, in the adult, about 2.5 cm in length. It consists of an outer, cartilaginous portion and an inner, bony portion, and is lined with skin. In the cartilaginous portion the skin lining contains hairs and sebaceous and ceruminous glands.

The middle-ear cavity, except for its lateral wall, the tympanic membrane, is surrounded by bone and lined with mucous membrane. It communicates through the aditus with the mastoid and through the Eustachian tube with the nasopharynx. It contains the three middle-ear ossicles—malleus, incus, and stapes—which connect the tympanic membrane with the oval window and represent the normal pathway for sound transmission across the middle ear, which acts as a mechanical transformer. The muscles are attached to the ossicular chain. The tensor tympani muscle attaches to the neck of the malleus, and the stapedius muscle attaches to the neck of the stapes. These muscles represent a protective mechanism for the inner ear against very intense sound.

The Eustachian tube measures about 36 mm in length and consists of a lateral, bony portion and a medial, cartilaginous portion. Lined by respiratory epithelium and opening into the lateral wall of the nasopharynx, it provides exchange of air to the middle ear by the brief opening action of the tensor and levator muscles of the palate during swallowing.

Located in the petrous portion of the temporal bone, enclosed in the otic capsule, the inner ear houses the membranous cochlea and vestibular labyrinth, the sense organs of hearing and balance. The membranous cochlea and the vestibular labyrinth consist of a system of epithelial-formed spaces and tubes containing endolymph. This system is surrounded by the perilymph-filled periotic labyrinth, which in turn is enclosed in the bony labyrinth of the otic capsule. The perilymphatic system communicates through the cochlear aqueduct with the subarachnoid space. The endolymphatic system of the cochlea communicates with the saccular labyrinth through the ductus reuniens. The cochlea resembles a snail shell in appearance, with two and three-quarter turns in a horizontal plane, with its lower basal end forming the medial wall (promontory) of the middle ear. Each turn is made up of three compartments. The upper compartment of the cochlea, the scala vestibuli, is associated with the oval window, while the lower compartment, the scala tympani, ends at the round window. Both scalae are perilymph-filled and communicate at the apex, through the helicotrema. Between the two is the medial compartment, the scala media, or the cochlear duct. It extends from the cochlear recess of the vestibule to the cupular cecum at the apex of the cochlea. Near its basal end, the ductus reuniens provides communications of endolymph with the saccule. The cochlear duct, on a transverse section, has a triangular form. The floor of the cochlear duct is formed by the rigid basilar membrane; the lateral wall, by the stria vascularis of the spiral ligament: with the vestibular, or Reissner's, membrane forming the third wall.

The sensory receptors and supporting structures responsive to acoustic energy are located along the basilar membrane and form the organ of Corti. The organ of Corti contains a single row of 3,500 inner hair cells and three to four rows of over 20,000 outer hair cells. These hair cells are the sensory receptors to acoustic stimuli. The upper surface of the hair cells is formed by a thickened cuticular plate in which the stereocilia hairs are embedded in a specific pattern. The major support for the hair cells is provided at the superior surface, where they are attached to the rigid reticular lamina, which in turn is firmly attached to the basilar membrane by means of the pillar cells. The tectorial membrane, in which the stereocilia of the hair cells are embedded, on the other hand, has a loose attachment to the basilar membrane. Thus, displacement of the basilar membrane will differentially influence the cell bodies of the hair cells and the overlying tectorial membrane. Movement of the basilar membrane, by bending the stereocilia of the hair cells, initiates a transduction response in the sensory receptors. Beneath each hair cell, terminations of the afferent eight nerve fibers extend through the tunnel of Corti and enter the osseous spiral lamina. From there, they continue through Rosenthal's canal into the cochlear modiolus to join the main body of the cochlear nerve. Sound reaching the ear initiates a chain of mechanical and neural events that result in volleys of nerve impulses in the cochlear nerve. The volleys are relayed over afferent pathways to cell groups in the pons, midbrain, thalamus, and auditory receiving areas of the cortex.

The membranous vestibular labyrinth consists of the saccule, utricle, the three (superior, lateral, and posterior) semicircular canals, and the endolymphatic duct and sac. It is filled with endolymph and surrounded with perilymph. Both ends of each semicircular canal open into the utricle. Near the utricle, each canal enlarges to form the ampulla, which houses the crista, which is covered by a specialized neural epithelium. Two types of hair cells are found in this structure. The stereocilia of hairs of these cells insert into a gelatinous cupula, which extends from the surface of the crista to the roof of the ampulla. Displacement of endolymphatic fluid within the semicircular canal deviates the cupula relative to the surface of the crista, thus bending the hairs and initiating a transduction response in the sensory receptors. Each semicircular canal, which lies in a plane at a right angle to the other two, responds to angular acceleration (rotation) in its own plane. The utricle occupies the elliptical recess; the saccule, the spherical recess of the vestibule. Their sensory receptors and supporting structures form the maculae; in the saccule, the macula lies in a vertical plane, perpendicular to the macula of the utricle. The receptor cells in each are hair cells of two kinds, with kinocilia and stereocilia hairs projecting into an overlying otolithic membrane. Pressure changes on the underlying sensory receptors initiate the transducing response in these hair cells. Utricle and saccule are responsive to positional changes of the body, with the utricle being stimulated by centrifugal and vertical, and the saccule stimulated by linear, acceleration. The nerve fibers from the semicircular canals, utricle, and saccule join to form the vestibular portion of the eighth cranial nerve to reach the vestibular nuclei in the floor of the fourth ventricle. These nuclei have connections with the third, fourth, sixth, and tenth cranial nerves, with the spinal cord and with the cerebellar cortex.

Whereas the middle ear derives its blood supply mainly from branches of the external carotid artery, the inner ear is supplied by a branch of either the basilar or anterior inferior cerebellar artery of the vertebral system.

Plate 34.
The Ear—Apparatus of Hearing and Equilibrium

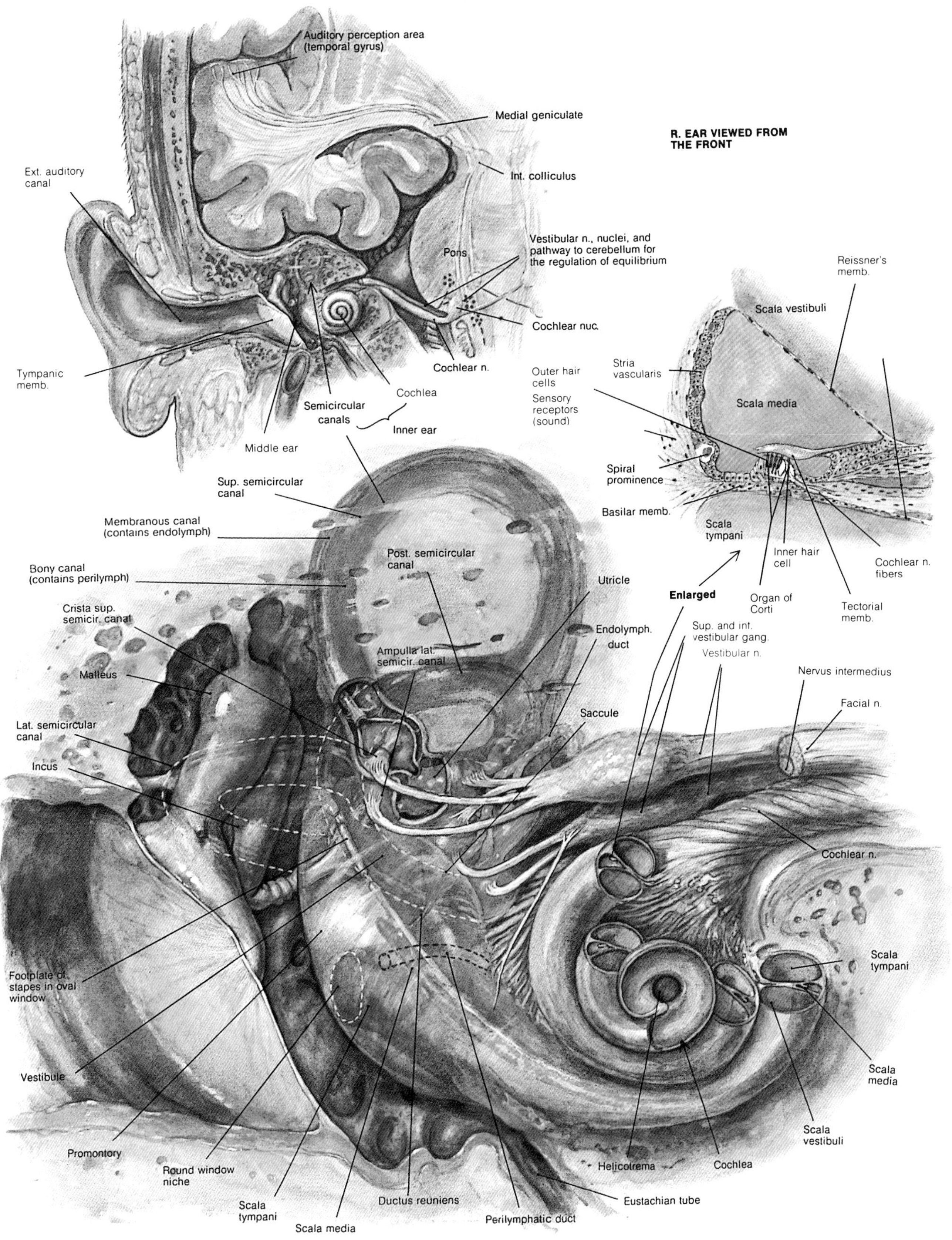

The Nose, Paranasal Sinuses, Pharynx, and Larynx

14

Margaret M. Fletcher, M.D.

The nose represents the superiormost portion of the upper respiratory system. It plays an important part in the conditioning of the inspired air for the lower respiratory tract. This conditioning includes (1) the control of temperature, (2) the control of humidity, and (3) the elimination of dust and infectious organisms. Variation of outside temperature is considerable, ranging from plus 100 to minus 40 degrees Fahrenheit, but regardless of the outside level, the temperature of the inspired air is converted to about body temperature by the time it reaches the nasopharynx. This is achieved during its brief passage, about one-fourth of a second, through the nose, and is accomplished by the extensive capillary bed in the erectile tissue of the nasal turbinates. The rapid expansion and contraction of this tissue and its large vascular spaces with a rapid blood flow guarantee an instantaneous heat transfer from blood to air or vice versa. Similarly, the humidity of the outside air reveals a wide range, varying from less than 1 percent to more than 90 percent. By the time the inspired air reaches the nasopharynx, its relative humidity is converted to a constant 75 to 80 percent. The required rapid transfer of water from the nasal mucosa to the inspired air or vice versa is provided by the mucous blanket which covers the entire nasal mucosa.

The third important function of the nose is the active elimination of dust, infective organisms, and other particulate matter from the air before it reaches the nasopharynx. These particles are deposited on the mucous blanket. Constant ciliary action carries these particles to the nasopharynx and pharynx, from whence they drain into the stomach. The nasal secretions contain, among other substances, immunoglobulin or lysozyme, an enzyme that destroys bacteria on contact. Most organisms that enter the nose are destroyed in this manner. The rest are eliminated by the hydrochloric acid and other gastric secretions in the stomach. The mucous blanket in the nose travels approximately five to ten millimeters per minute. It is replaced about every twenty minutes from the submucosal glands and goblet cells. Thus, one of the several protective functions of the mucous blanket is to provide an effective defense mechanism against infections.

INTERNAL NOSE

The anterior nares open into a dilated portion of the nasal chamber called the *vestibule.* The vestibule is lined with skin that bears stiff hairs, which serve to filter large particles from the inspired air.

The surface area of the internal nasal chamber is doubled by the scroll-like projections on the lateral wall. These are the superior, middle, and inferior turbinates. Groovelike passages, called the *superior, middle,* and *inferior meatus,* exist between the turbinates. Beneath the inferior turbinate the nasal lacrimal duct enters the inferior meatus, 2 cm behind the mucocutaneous junction. The frontal, maxillary, and ethmoid sinuses open into the middle meatus beneath the middle turbinate.

OLFACTION

The olfactory epithelium covers the superior turbinate and adjacent nasal septum. Olfactory cells are bipolar neurons that are stimulated by lipid soluble substances in the inspired air. The axons penetrate the cribriform plate and synapse in the olfactory bulb. Secondary neurons then synapse in the central nervous system.

Plate 35.
The Nose and Paranasal Sinuses

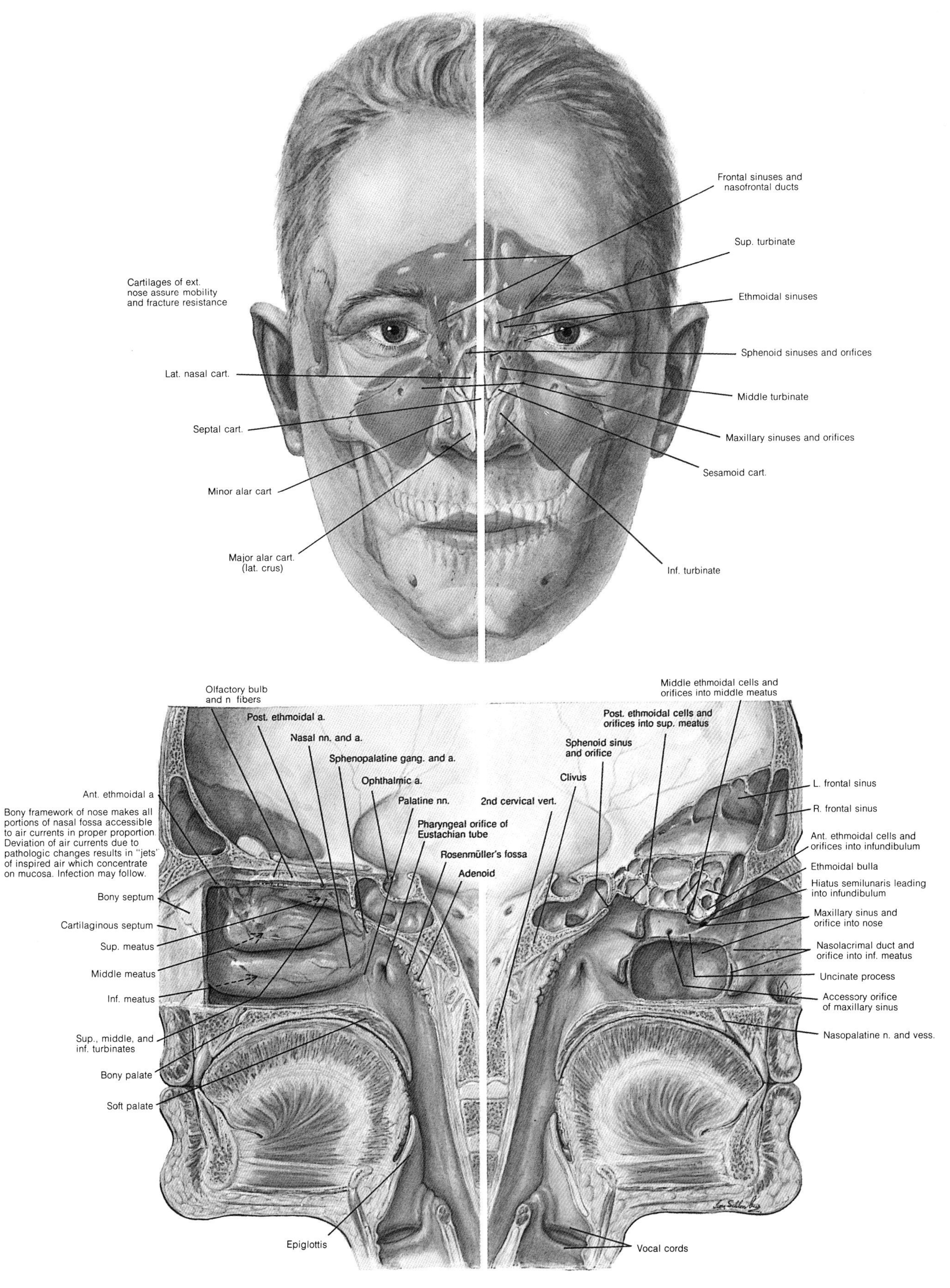

PARANASAL SINUSES

The paranasal sinuses are pneumatized areas in the frontal, maxillary, ethmoid, and sphenoid bone. They are lined with respiratory epithelium, which is continuous with that of the nasal cavity or upper airway. While the function of the paranasal sinuses is not entirely clear, they are important in the production of mucus and antibodies necessary for resistance to upper respiratory infections. They serve as resonating chambers and decrease the weight of the skull bones.

The frontal sinuses are located in the frontal bone. They are separated by a midline septum and extend laterally behind the superciliary arch and deeply over the orbit. The ethmoid sinuses are located between the eyes. They consist of a labyrinth of a number of air cells (3 to 18). Their boundaries are completed by the frontal, lacrimal, sphenoid, maxillary, and palatine bones. The ethmoid air cells open into the semilunar hiatus of the middle meatus.

The sphenoid sinus is important because of the company it keeps. It is bound laterally by the cavernous sinuses, which contain the carotid artery and third, fourth, fifth, and sixth cranial nerves. The superior boundary of the sphenoid sinus contains the optic canal, dura mater, and the pituitary gland. The sphenoid sinus empties into the sphenoethmoidal recess immediately lateral to the attachment of the nasal septum in the superior meatus.

The maxillary sinus is the largest of the paranasal sinuses. Its floor is formed by the alveolar process of the maxilla. A very thin plate of bone exists between the tooth roots and the sinus, a frequent source for infection. Laterally, the maxillary sinus extends into the zygomatic arch. The roof of the sinus is also the floor of the orbit.

PHARYNX

The pharynx serves as a common chamber for the respiratory and digestive tracts. It is divided into nasal, oral, and laryngeal portions and is lined by respiratory epithelium superiorly and squamous epithelium inferiorly. It contains important lymphoid tissue for further protection of the human organism. The constrictor muscles of the pharyngeal wall are circular muscle fibers, which are important in swallowing.

NASOPHARYNX

The nasopharynx is bounded superiorly by the base of the skull and sphenoid rostrum, posteriorly by the cervical vertebrae, anteriorly by the posterior choana of the nose. The lateral walls contain the Eustachian tube orifices and the cartilagenous torus tubarius. Inferiorly it opens into the oral pharynx at the level of the soft palate. The Eustachian tube is important for ventilation of the middle ear, air being necessary in the middle ear in order to transmit sound through the ossicular chain. The pharyngeal tonsils on the posterior wall are important in the production of antibodies.

ORAL PHARYNX

From the level of the soft palate, the oral pharynx represents the digestive entrance of the chamber. The lateral walls are occupied by the palatine tonsils, which are bounded by the palatoglossal and palatopharyngeal folds. Inferiorly it extends to the level of the hyoid bone.

LARYNGEAL PHARYNX

The laryngeal pharynx lies behind the vestibule and posterior commissure of the larynx. Its anterior wall consists of the epiglottis, aryepiglottic folds, posterior commissure of the larynx, piriform sinuses laterally, and the posterior pharyngeal wall.

SWALLOWING REFLEX

A bolus of food is propelled posteriorly by the tongue. The nasopharynx is closed superiorly. As the food passes into the vallecula, the epiglottis folds over the laryngeal vestibule and the laryngeal muscles close reflexively. The cricopharyngeal muscle relaxes and the bolus of food enters the esophagus.

LARYNX

The larynx receives inspired air, which has been warmed, humidified, and filtered, and passes it on through the trachea and the bronchi to the lungs. The larynx acts as a valve to prevent the passage of foods and liquid into the area. There are nine laryngeal cartilages, which give the larynx a rigid form. Muscles act on the cartilages to modify the laryngeal aperture. The three single cartilages are the epiglottis, thyroid, and cricoid. The thyroid cartilage is composed of two lamina and provides support for the glottis. The strap muscles and inferior constrictor muscles of the pharynx are attached to the thyroid cartilage.

VOCAL APPARATUS

The form of the larynx is modified to control the expulsion of air from the lungs in order to produce sound. Paired arytenoid cartilages rotate on the cricoid cartilage to change the length and tension of the vocal cord for the production of sound.

The glottis is opened during inspiration by the posterior cricoarytenoid muscles. The glottis is closed during phonation and also to protect the airway by contraction of the lateral cricoarytenoid, transverse arytenoid, and thyroarytenoid muscles.

Plate 36.
The Pharynx and Larynx

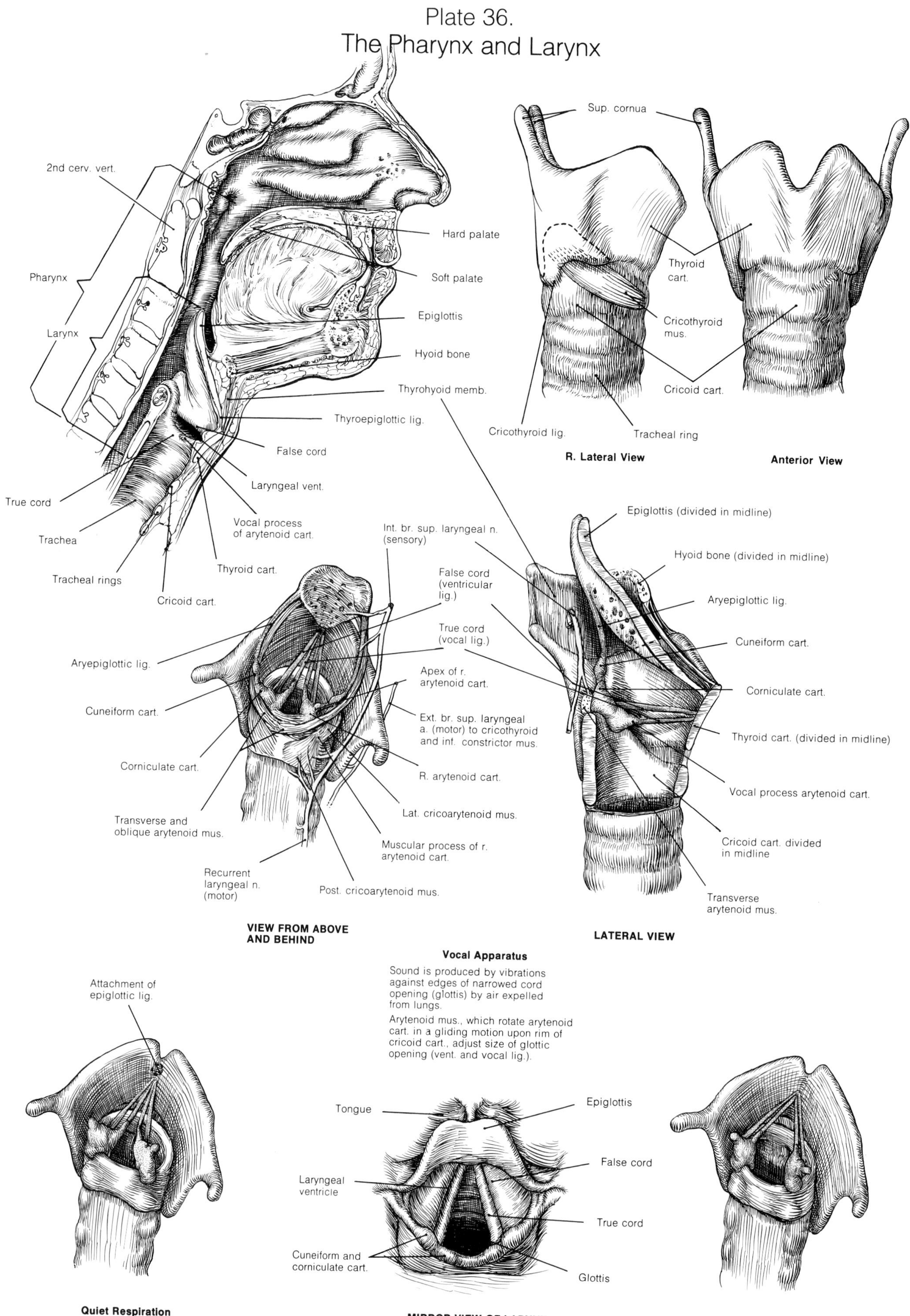

The Head and Neck

Melvin H. Epstein, M.D.
Donald S. Gann, M.D.
David W. Heese, D.D.S.
James J. Ryan, M.D.

The bones of the head consist of the cranium, encompassing the brain, and the bones of the facial skeleton. The facial bones are conveniently considered as bones of the midface—those of the orbit, nose, zygoma (cheek), and maxilla (upper jaw). The single bone of the lower face is the mandible, the only movable facial bone. The muscles associated with the midface are the muscles of facial expression, which, taking origin from the bones of the facial skeleton and attaching to the soft tissues of the eyelids, nose, cheeks, and lips, produce the facial movements of emotion and expression. These muscles are all innervated by the seventh cranial nerve. The muscles attached to the mandible, the muscles of mastication, arise from the cranium (i.e., the temporalis muscle) and the facial bones (i.e., the masseter muscle) and, with the tongue, form the floor of the mouth. These are all innervated by the fifth cranial nerve. In the neck the most prominent cervical muscle, the sternomastoid, divides the neck structures into anterior and posterior triangles.

BLOOD SUPPLY OF THE HEAD AND NECK

The scalp, face, and neck derive their blood supply primarily from the external carotid arteries. The external carotid is a terminal branch of the common carotid, which begins in the carotid triangle and passes upward and medial to the digastric muscle and the stylohyoid muscles. Eight branches arise from the external carotid artery. The superior thyroid artery arises at its lowest portion and terminates in the thyroid gland and the adjacent muscles and membranes attached to the thyroid cartilage. The lingual branch originates opposite the hyoid bone and ends beneath the tip of the tongue. The facial artery ends at the medial angle of the eye, where it anastomoses with the terminal branch of the ophthalmic artery. It provides the primary blood supply to the face, the tonsils, and the muscles at the base of the skull. The occipital branch supplies blood to the muscles of the back of the neck, in addition to the auricle and the posterior part of the scalp. There are meningeal branches that enter the skull and supply the dura mater. The posterior auricular branch ends near the mastoid process, behind the auricle. This artery supplies blood to the tympanic cavity, the antrum, the vestibule, the semicircular canals, and the mastoid air cells, as well as the scalp in the posterior temporal region.

The ascending pharyngeal artery supplies the wall of the pharynx and the soft palate. The superficial temporal artery, one of the terminal branches of the external carotid, has a small parotid branch and a small auricular branch, and the remaining branches give blood to the scalp, including the frontal belly of the occipitofrontalis and orbicularis oculi muscles. The maxillary artery is divided into three parts: the first part furnishes blood to the dura mater, the mandibular joint, the tympanic cavity, the mandible, teeth, and gingiva; the second part supplies the masseter muscle, the temporal muscle, and the buccal mucosa; and the third part furnishes blood to the orbit, the roof of the mouth, the gums and mucous membranes of the hard palate, the pharynx, the roof of the nose, the sphenoid sinus, and the ethmoid sinuses.

The vertebral artery is the first branch from the subclavian artery. It ascends, giving off muscular branches to the deep muscles of the neck and the suboccipital muscles, and ultimately enters the cranial cavity to supply the brain. The thyrocervical trunk, arising from the subclavian system, primarily supplies the muscles in the anterior part of the neck. The deep cervical artery also arises from the subclavian and primarily supplies the posterior muscles of the neck.

NERVE SUPPLY OF THE HEAD AND NECK

The anterior and posterior nerve roots divide into anterior and posterior primary rami, each of which are mixed sensory and motor nerves. The first cervical nerve is small. The posterior ramus supplies some of the muscles in the back of the neck, and the anterior primary ramus joins the hypoglossal nerve to supply the muscles attached to the larynx and hyoid bone. The posterior primary ramus of the second cervical nerve becomes the greater occipital nerve, the chief cutaneous nerve for the posterior part of the head. The posterior primary rami of the third, fourth, fifth, and sixth cervical nerves, small branches of minor importance, supply the skin of the back of the neck and some of the paraspinal muscles. The anterior primary rami of the first four cervical nerves form the cervical plexus. The cutaneous branches include the lesser occipital, greater auricular, and anterior cutaneous nerve from C2 and C3 roots, innervating the lateral skin of the neck and the scalp behind the ear. The supraclavicular nerves from the C3 and C4 root innervate the inferior lateral neck and the area of the chest several centimeters below the clavicle. There are muscular branches that arise from C2, C3, and C4 that innervate the sternomastoid, trapezius, and scalene muscles, and there are muscular branches that arise from C1 through C4 that innervate the prevertebral muscles and the diaphragm. There are also communicating branches that join the accessory nerve in its course to the trapezius muscle and communicating branches to the vagus nerve, hypoglossal nerve, and ansa hypoglossi.

The face is innervated by the fifth cranial nerve, which is divided into three, the first division innervating the eye and some of the structures of the orbit, the second the cheek, and the third the chin. The nerve is discussed in detail in another section.

LYMPHATIC SYSTEM OF THE HEAD AND NECK

The lymphatic system of the head and neck converges principally about the jugular veins in the carotid sheaths, deep to the sternomastoid muscle. The superficial lymph nodes of the head (the occipital nodes, posterior auricular nodes, anterior auricular nodes, and superficial parotid nodes) drain into the superior deep cervical nodes. Likewise, the superficial lymph nodes of the neck (the submental and submaxillary nodes) have a similar drainage. The deep lymph nodes of the neck are the retropharyngeal nodes medially, the juxtavisceral nodes (infrahyoid, prelaryngeal, pretracheal, and paratracheal nodes) anteriorly, and the superior medial and superior lateral nodes. All converge to drain into the supraclavicular nodes and thence down into the mediastinum. In addition to carrying lymphatic fluid from these regions, these nodes are active in body defenses against infection and cancer, in both of which states enlargement is common.

The other major glandular system of the head includes the major salivary glands. The largest of these, occupying the lateral cheek, is the parotid gland. In the floor of the mouth lie the small submaxillary glands and the smaller sublingual glands. These, in association with the innumerable minor salivary glands of the mouth, generate the salivary secretion commencing the digestive process. This varies greatly in volume, depending on digestive stimuli, but can amount to more than two liters daily.

THE ORAL CAVITY

The oral cavity is the beginning of the digestive tract. The bony support consists superiorly of the maxilla and inferiorly of the mobile mandible, attached by the temporomandibular joint. The cavity is divided into the vestibule, bounded medially by the teeth, and the cavity proper, bounded laterally and anteriorly by the teeth, posteriorly by the oral pharynx, superiorly by the hard and soft palate, and inferiorly by the attachment muscles of the tongue and the mylohyoid and genioglossus muscles.

Innervation of the oral cavity is supplied by the trigeminal nerve, whose branches are sensory to the teeth, mucous membranes, and anterior two-thirds of the tongue; and motor to the five muscles of mastication (masseter, temporalis, medial, and lateral pterygoid, and the anterior belly of the digastric). Other motor supply about the oral cavity comes from the facial nerve.

Blood supply to the oral cavity comes from various branches of the external carotid artery.

The lining of the area is mucous membrane, which contains many minor salivary glands. Surrounding the cavity and exiting into it are the major salivary glands: the parotid (Stensen's duct), the submaxillary (Wharton's duct), and the sublingual glands. These glands provide moisture for lubrication and add digestive enzymes, which begin the breakdown of the food.

Most of the cavity is filled by the highly muscular tongue, covered in the anterior two-thirds by hairy filiform and flat fungiform papillae. The posterior one-third is covered with the specialized circumvallate papillae. The tongue functions to move food to the grinding surfaces of the teeth and finally to deliver the bolus posteriorly to the oral pharynx and the esophagus. The oral pharynx is isolated from the nasal pharynx at this time by the lifting action of the soft palate.

The unique tissues in the mouth are the teeth, which are supported in the alveolar processes of the maxilla and mandible. In the adult dentition there are thirty-two teeth, divided into specialized tooth forms: the central and lateral incisors for cutting, the canines for tearing, the two premolars (bicuspids), and the three molars for

Plate 37.
Innervation of the Teeth and Facial Muscles; Salivary Glands and Muscles of Mastication

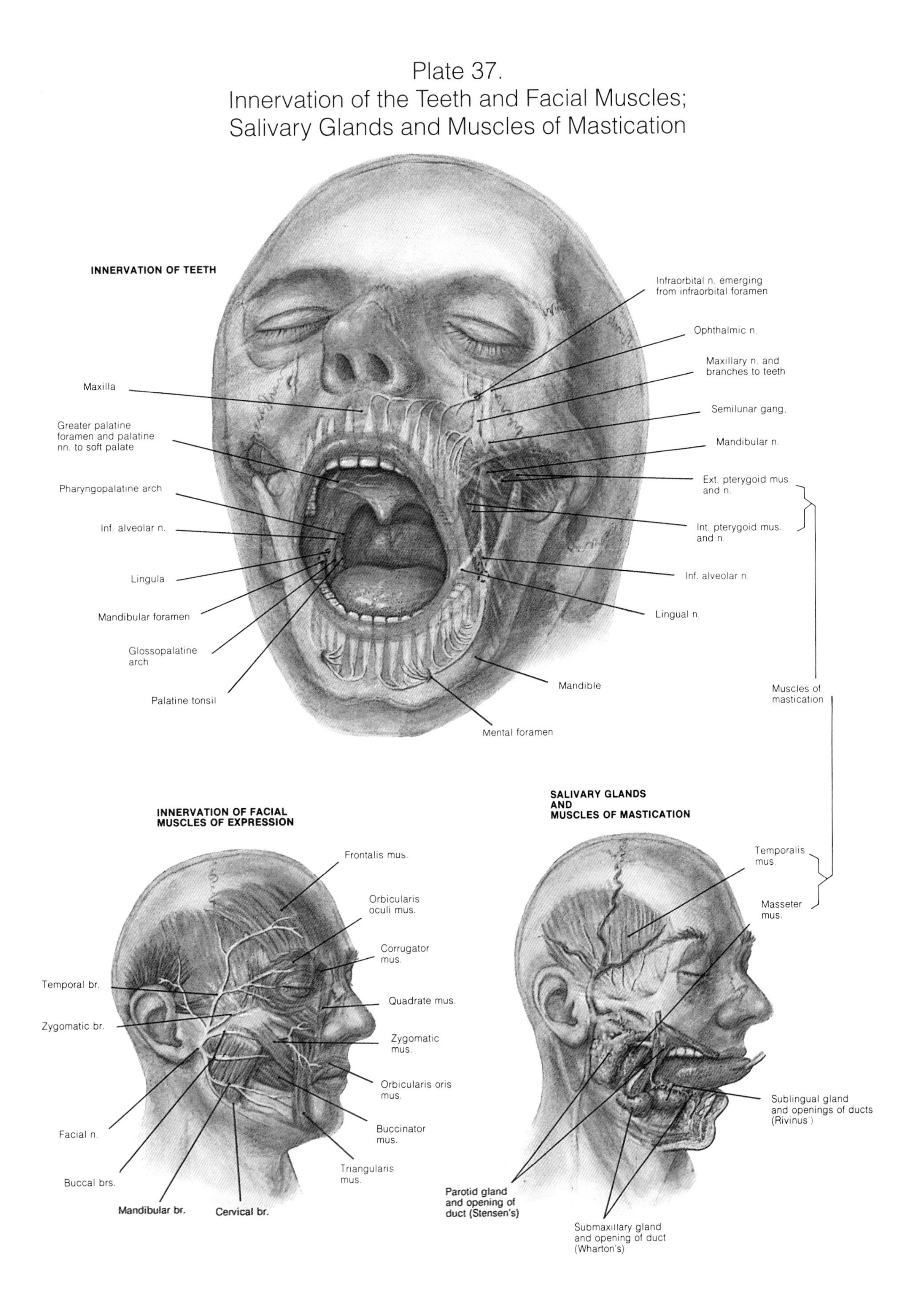

grinding. Teeth are ectodermal (enamel) and mesodermal (dentin and pulp) in origin, and unlike most other body tissues are not capable of repair. Deciduous teeth are similarly named, but instead of premolars there are but two deciduous molars, which are replaced by premolars of the adult dentition. The permanent molars do not have deciduous predecessors.

THYROID AND PARATHYROID

The thyroid secretes hormones with diverse metabolic effects that act on nearly every tissue in the body. The thyroid follicles synthesize thyroxin (T4) and triiodothyronine (T3) from iodine and tyrosine. The uptake of iodide into the thyroid gland and its subsequent organification is under the control of the thyroid-stimulating hormone (TSH) from the anterior pituitary. TSH also controls release of a thyroid hormone in the bloodstream. Although both T4 and T3 are secreted, it is currently thought that most peripheral actions of the hormones occur through T3 following peripheral conversion from T4, the more slowly metabolized hormone. However, the feedback control of TSH release by the anterior pituitary operates through both T3 and T4. This feedback mechanism provides a control mechanism so that the principal mediator of release of TSH by the pituitary is the peripheral utilization of the thyroid hormones. The set point for this control system is provided by thyrotropin-releasing hormone (TRH) for the hypothalamus. This tripeptide provides a basis for some aspects of physiologic regulation. For example, cold leads to increased secretion of TRH. Most stresses, including surgery, decrease secretion of TSH, presumably through decreases in TRH. Although adrenal steroids also reduce TSH, this effect is also seen after adrenalectomy.

The thyroid hormones lead to increased oxygen consumption in most tissues except brain, spleen, and testes. The hormones lead directly to increased incorporation of amino acids into all tissues except for those named above, and lead to induction of a variety of enzymes. This induction of enzymes may be responsible for the increased metabolic rate seen with the hormones. Release of free fatty acids and glycogenolysis are enhanced by increased sensitivity to catecholamines. Finally, with the increased metabolic rate, the utilization of various coenzymes and vitamins is increased.

In addition, the C cells of the thyroid secrete a hormone, thyrocalcitonin, which is entirely independent of the system controlling T3 and T4. Secretion of thyrocalcitonin is stimulated by a rise in serum calcium and inhibited by a fall in the ion. This hormone thus participates, together with parathyroid hormone, in a double negative feedback system controlling serum calcium concentration. The hormone acts on bone to increase calcium uptake and bone formation and to decrease activity of osteoclasts. It also decreases renal excretion of calcium.

PARATHYROID

The parathyroid glands are the source of parathormone (PTH), which is the principal hormone controlling calcium metabolism. There are commonly four parathyroid glands, all of which receive their blood supply from branches of the inferior thyroid artery. The superior parathyroids may also receive branches from the superior thyroid arteries. In addition, the inferior parathyroid glands commonly lie anterior to the recurrent laryngeal nerve, whereas the superior parathyroid glands commonly lie posterior to this nerve. This relationship and the consistent arterial supply facilitate location of the parathyroid glands at surgery.

The major action of PTH is to increase serum calcium concentration by mobilizing calcium from bone. The action of the hormone appears to depend upon activation of cyclic AMP and perhaps on increased entry of magnesium ion into cells. In addition, osteoclastic activity is enhanced. PTH also acts upon the kidney to inhibit reabsorption of phosphate in the proximal tubule. The hormone also increases renal reabsorption of calcium, but because filtration of calcium is enhanced, there is a net increase in calcium excretion. Magnesium excretion is also enhanced, again because the filtered load is increased secondary to bone reabsorption. Finally, PTH acts upon the gut to increase absorption of calcium, but vitamin D is required for this effect.

Plate 38.
Sagittal Section of the Head and Neck; Thyroid and Parathyroid Glands

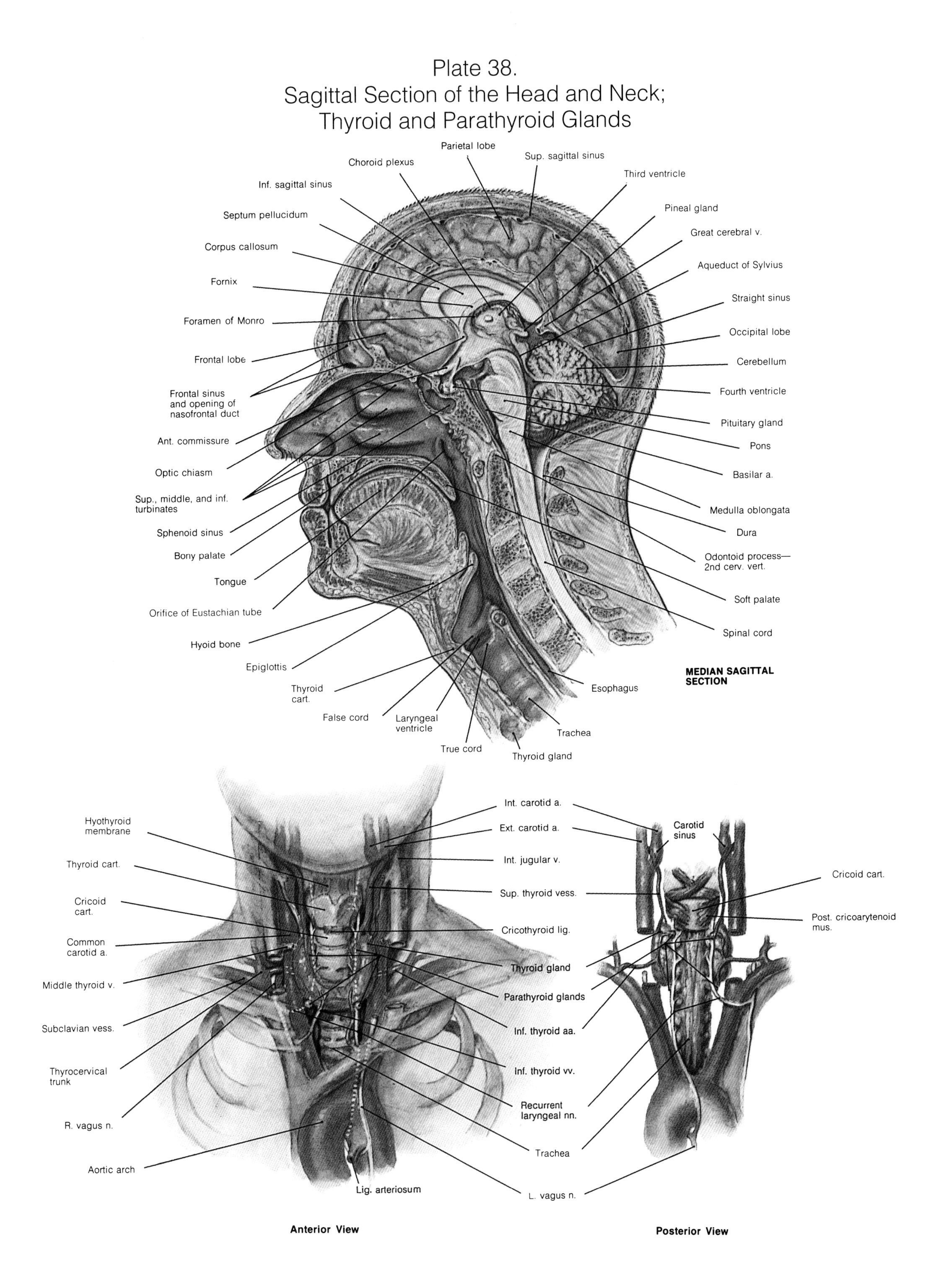

The Endocrine Glands

Donald S. Gann, M.D.

PITUITARY GLAND

The pituitary gland is attached to the median eminence of the hypothalamus. In addition to a direct arterial supply, the anterior lobe of this gland receives blood through a portal system from the hypothalamus. The activity of the posterior lobe is controlled directly by nerves with their cell bodies in supraoptic and paraventricular nuclei of the hypothalamus. This lobe secretes the hormones vasopressin and oxytocin. Secretion of vasopressin occurs in response to hypovolemia and to increased serum osmolality. Vasopressin increases reabsorption of water by the kidney by increasing permeability of the distal nephron. Increased secretion of oxytocin, which increases the letdown of milk and uterine tension, follows sucking or uterine distention.

Secretion of the hormones of the anterior pituitary gland is under the control of neurohormonal-stimulating and inhibiting factors released from the hypothalamus into the hypophysial portal vessels. The principal hormones include corticotropin (ACTH), melanophore-stimulating hormone, thyrotropin, prolactin, growth hormone, and gonadotropins. For each of these substances there appears to be a specific releasing hormone, and in some cases an inhibitory hormone as well. The principal pituitary hormones come from at least five different cell types in the human pituitary gland.

ACTH is released in response to all forms of stress, including pain, trauma, hemorrhage, fear, general sensory stimulation, cold, hypoglycemia, and immobilization. It exerts its principal effect on the zona fasciculata of the adrenal cortex to elicit secretion of cortisol. It also has a transient stimulating effect on secretion of aldosterone. In addition, ACTH acts independent of the adrenal to stimulate lipolysis.

Melanocyte-stimulating hormone (MSH) is a hormone that shares an amino acid sequence with amino acids 4 to 10 of ACTH. It is a major hormone in fish and amphibians, and functions in man to cause darkening of the skin. Little is known about its control. Its secretion is inhibited by MSH-inhibiting factor from the hypothalamus.

Growth hormone has a wide variety of metabolic effects. It increases incorporation of amino acids in muscle and liver and stimulates growth of cartilaginous epiphyses, which effects probably account for its growth-stimulating action. In addition, the hormone leads to retention of sodium, potassium, calcium, and phosphate by the kidney. Growth hormone has a transient lipogenic effect by increasing uptake of glucose into fat cells, but this gives way to a predominant lipolytic effect. In addition, growth hormone inhibits glucose entry into muscle cells and ultimately into fat cells, thus antagonizing the action of insulin. Gluconeogenesis is increased, adding further to the increase in concentration of glucose in blood.

The thyroid-stimulating hormone (TSH) acts on the thyroid gland to stimulate synthesis and secretion of the thyroid hormones thyroxin and triiodothyronine. Release of TSH is the result of an interaction between hypothalamic thyrotropin-releasing hormone (TRH), which stimulates release of TSH, and the feedback effects of thyroxin and triiodothyronine, which inhibit that release. In general the system acts to maintain a constant level of thyroid hormones in blood, so that its major drive is from peripheral utilization of hormone. Cold increases release of TRH, whereas general stress appears to decrease release of TRH.

Prolactin is a hormone that acts to maintain both lactation and the stability of the ovarian corpus luteum. Its secretion is controlled at least in part by TRH. There are some apparent interactions with circulating estrogens and progestins, but the mechanisms of control are not clear. In addition, there is some evidence that prolactin may be active in the control of renal handling of salt and water. Again, the physiologic mechanisms that may underlie this effect are obscure.

Follicle-stimulating hormone (FSH) facilitates the development of ovarian follicles and testicular tubules. Its secretion is controlled at least in part by circulating levels of gonadal steroids. Release of FSH is usually diminished after surgical stress.

Luteinizing hormone (LH) controls ovulation, the development of the corpus luteum, and, to some extent, estrogen secretion in the female, and stimulates secretion of testosterone in the male. Its release is inhibited by estrogen and by testosterone except under a priming condition, in which case the estrogen can facilitate the release of LH, bringing about a surge which leads to ovulation.

ADRENALS

The adrenal glands, as suggested by their name, lie above the kidneys bilaterally. They receive their arterial supply through end arteries that arise from phrenic, aortic, and renal vessels. In contrast, there is a central vein that collects all venous drainage from the adrenal cortex and medulla. The adrenal cortex consists of three zones and appears to be under humoral control. In contrast, the adrenal medulla consists of a single zone and is controlled primarily by the sympathetic nervous system.

The most external zone of the adrenal cortex, the zona glomerulosa, is responsible for secretion of aldosterone. This hormone controls, in part, the renal handling of salt and water. In the distal renal tubule it facilitates reabsorption of sodium chloride. In the late distal tubule and collecting duct it facilitates exchange of sodium for potassium and hydrogen and thus is the major factor controlling excretion of potassium and hydrogen. Secretion of aldosterone is controlled primarily by angiotensin II, formed in response to release of the enzyme renin by the kidney. Its secretion is also controlled by ACTH acutely and by the circulating level of potassium ion. The release of renin is controlled in a complex manner, but the cardiovascular baroreceptors and the sympathetic nervous system appear to play a dominant role.

The principal hormone secreted by the middle zone of the adrenal cortex, the zona fasciculata, is cortisol. Cortisol has a variety of metabolic effects. In all tissues except liver it inhibits incorporation of amino acids into cells. This leads to an antianabolic state and to net tissue catabolism. It may also be responsible for the anti-inflammatory action of this steroid, since lymphoid tissues are particularly sensitive to its action. In the liver cortisol facilitates the uptake of amino acids and stimulates both gluconeogenesis and albumin synthesis. In all cells cortisol inhibits the uptake of glucose. These effects lead to a net increase in serum glucose concentration. In addition, the hormone facilitates the movement of potassium and water from cells into the extracellular space. Except for the immediate action on glucose uptake, most of the actions of the steroid appear to involve incorporation of the hormone into the nucleus and formation of mRNA. The control of cortisol secretion is almost exclusively by pituitary ACTH.

The zona reticularis of the adrenal gland, the inner cortical zone, is responsible for adrenal secretion of androgens and estrogens. This zone is primarily controlled by ACTH, although FSH or other pituitary factors may also play a role. Substantial amounts of sex steroids may be of adrenal origin.

The adrenal medulla is controlled primarily by the sympathetic nervous system. The transmitter is acetylcholine, released by sympathetic nerve endings and stimulating directly synthesis and release of epinephrine and norepinephrine. These hormones act to stimulate α and β receptors, leading to peripheral vasoconstriction, increased myocardial contractility, glycogenolysis, lipolysis, and a multitude of other effects. The release of catecholamines by the adrenal medulla is also facilitated by angiotensin II, so that there is a positive feedback mechanism relating these two factors. Finally, the synthesis of epinephrine requires the adrenal cortical hormone cortisol.

PANCREAS

The endocrine functions of the pancreas are based in the islands of Langerhans. These islands are distributed throughout the pancreatic tissue but concentrated somewhat in the tail of the pancreas. The principal functions of the island cells are secretion of insulin (β-cell) and of glucagon (α-cell). Insulin facilitates glucose transport into cells together with potassium, increases glycogenesis, and lowers blood concentration of glucose. In addition, insulin acts on muscle to increase incorporation of amino acids and to increase formation of protein. Insulin also acts on fat cells to increase glucose entry and oxidation and to lead to lipogenesis. Insulin does not affect glucose entry into liver cells but does increase glycogen formation, decrease glycolytic activity, and induce glucose-6-phosphate dehydrogenase activity. Secretion of insulin is stimulated by an increase in plasma glucose and by parasympathetic nervous activity, and is inhibited by sympathetic nervous activity and by epinephrine.

Glucagon increases glycogen breakdown and glycogenolysis in the liver but stimulates glucose uptake into muscle. It stimulates lipolysis but lowers plasma fatty acids, probably by increasing uptake into liver and muscle. It inhibits amino acid incorporation into muscle. In addition, this hormone decreases serum

Plate 39.
Endocrine Glands

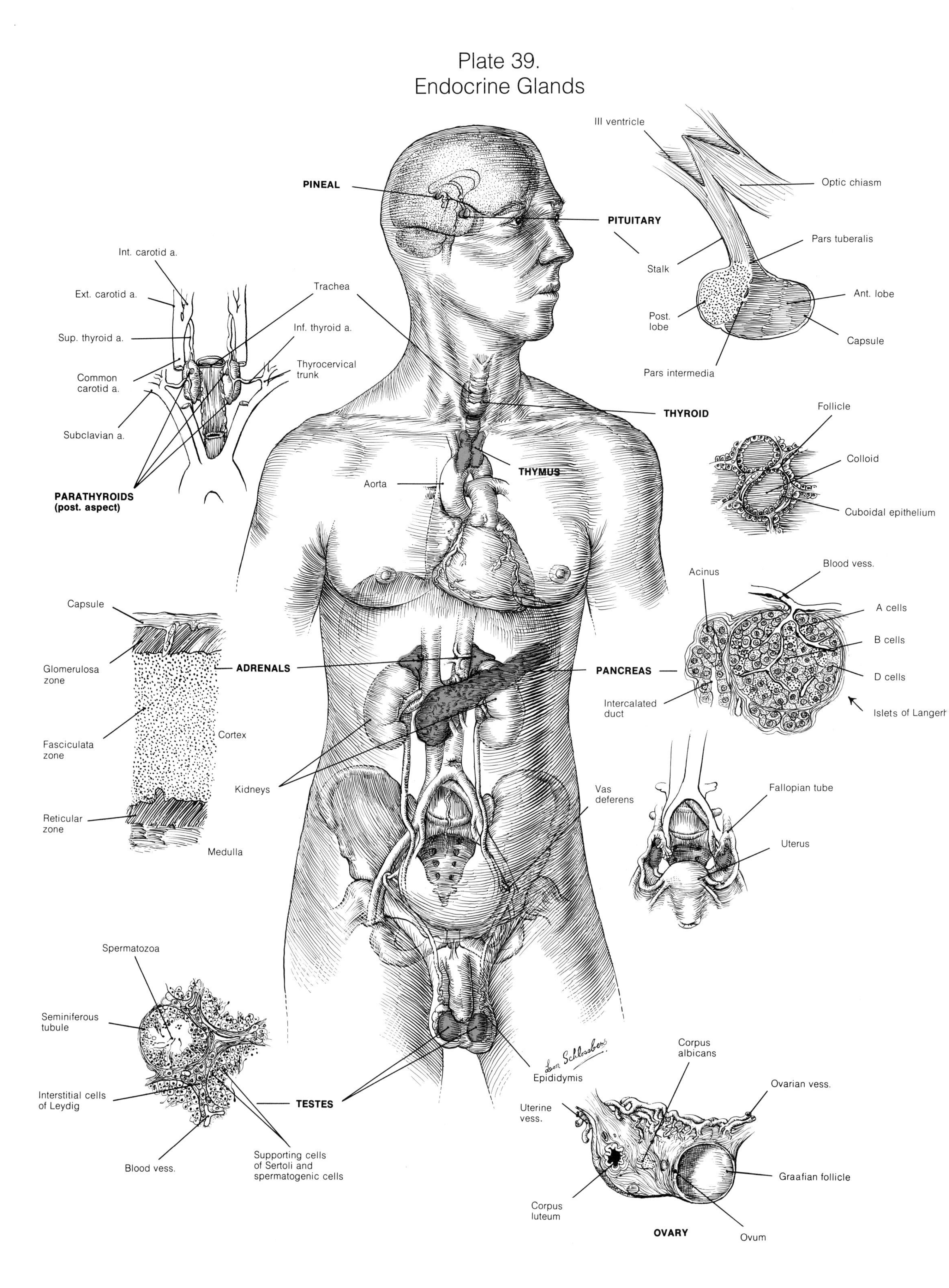

Plate 40.
Physiology of Endocrine Glands (Schematic Summary)

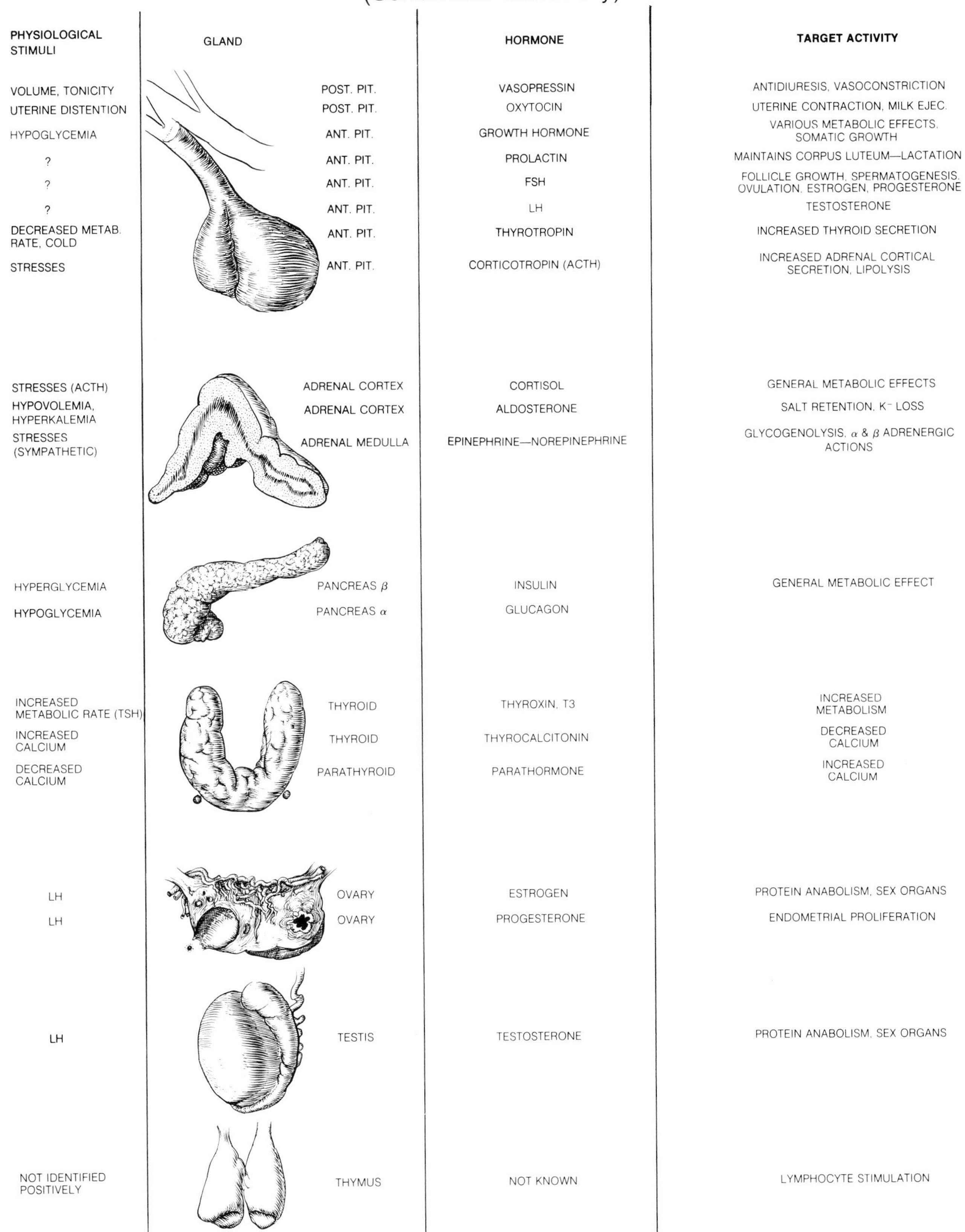

PHYSIOLOGICAL STIMULI	GLAND	HORMONE	TARGET ACTIVITY
VOLUME, TONICITY	POST. PIT.	VASOPRESSIN	ANTIDIURESIS, VASOCONSTRICTION
UTERINE DISTENTION	POST. PIT.	OXYTOCIN	UTERINE CONTRACTION, MILK EJEC.
HYPOGLYCEMIA	ANT. PIT.	GROWTH HORMONE	VARIOUS METABOLIC EFFECTS, SOMATIC GROWTH
?	ANT. PIT.	PROLACTIN	MAINTAINS CORPUS LUTEUM—LACTATION
?	ANT. PIT.	FSH	FOLLICLE GROWTH, SPERMATOGENESIS, OVULATION, ESTROGEN, PROGESTERONE
?	ANT. PIT.	LH	TESTOSTERONE
DECREASED METAB. RATE, COLD	ANT. PIT.	THYROTROPIN	INCREASED THYROID SECRETION
STRESSES	ANT. PIT.	CORTICOTROPIN (ACTH)	INCREASED ADRENAL CORTICAL SECRETION, LIPOLYSIS
STRESSES (ACTH)	ADRENAL CORTEX	CORTISOL	GENERAL METABOLIC EFFECTS
HYPOVOLEMIA, HYPERKALEMIA	ADRENAL CORTEX	ALDOSTERONE	SALT RETENTION, K^- LOSS
STRESSES (SYMPATHETIC)	ADRENAL MEDULLA	EPINEPHRINE—NOREPINEPHRINE	GLYCOGENOLYSIS, α & β ADRENERGIC ACTIONS
HYPERGLYCEMIA	PANCREAS β	INSULIN	GENERAL METABOLIC EFFECT
HYPOGLYCEMIA	PANCREAS α	GLUCAGON	
INCREASED METABOLIC RATE (TSH)	THYROID	THYROXIN, T3	INCREASED METABOLISM
INCREASED CALCIUM	THYROID	THYROCALCITONIN	DECREASED CALCIUM
DECREASED CALCIUM	PARATHYROID	PARATHORMONE	INCREASED CALCIUM
LH	OVARY	ESTROGEN	PROTEIN ANABOLISM, SEX ORGANS
LH	OVARY	PROGESTERONE	ENDOMETRIAL PROLIFERATION
LH	TESTIS	TESTOSTERONE	PROTEIN ANABOLISM, SEX ORGANS
NOT IDENTIFIED POSITIVELY	THYMUS	NOT KNOWN	LYMPHOCYTE STIMULATION

calcium and increases cardiac contractility. Secretion of glucagon is stimulated by hypoglycemia and by sympathetic nervous activity and inhibited by parasympathetic nervous activity.

THYROID

The thyroid gland is formed of two lobes, which lie over the trachea in the neck. Its principal secretory products are thyroxin and triiodothyronine, which regulate metabolism primarily through induction of enzymes. These hormones act at a number of sites synergistically with the catecholamines. In addition, both hormones act on the pituitary gland to inhibit release of TSH, which in turn stimulates the secretion of thyroid hormones. The thyroid also secretes thyrocalcitonin, which increases calcium incorporation into bone and decreases calcium excretion by the kidney. Secretion of this hormone is stimulated by hypercalcemia, and its net effect is to decrease the level of serum calcium in blood.

PARATHYROIDS

The parathyroid glands usually number four and are located on either side adjacent to the thyroid gland. Parathormone, the principal secretory product, acts to mobilize calcium from bone, increase gut absorption of calcium, and increase renal excretion of phosphate. These effects interact to increase the serum concentration of calcium and decrease the serum concentration of phosphate. Parathormone secretion is stimulated by hypocalcemia and inhibited by hypercalcemia.

TESTES

The principal secretory product of the testes is testosterone. This hormone is secreted by the interstitial cells in response to pituitary luteinizing hormone (LH). Testosterone acts to produce virilization and to increase amino acid incorporation into muscle, liver, and kidney. Bone growth is accelerated, but terminates when the epiphyses close in response to the action of the hormone. Red cell production is also stimulated. Some estrogen is also secreted by the testes in response to the action of FSH and LH.

THYMUS

The thymus lies in the anterior mediastinum anterior to the trachea. No specific hormones have been identified, but there appear to be one or more thymic hormones. The thymus appears to inhibit production of cells of the anterior pituitary that synthesize growth hormone, whereas growth hormone in turn increases thymic size. In addition, a thymic factor increases the number of stem cells in bone marrow. There is also evidence for a lymphocytosis-stimulating factor and for a factor that induces immunocompetence. In addition, a factor has been identified that inhibits neuromuscular transmission. Whether these effects are the result of actions of a single hormone or of multiple hormones is not known.

The Mediastinum and the Thymus Gland

Gregory B. Bulkley, M.D.

MEDIASTINUM

The mediastinum is the portion of the chest cavity that does not include the lungs and trachea. It is bordered by the thoracic inlet superiorly and the diaphragm inferiorly, the sternum anteriorly, the thoracic spine posteriorly, and the medial parietal thoracic pleurae laterally. It contains the heart, great vessels, and esophagus, the vagus (and recurrent laryngeal) and phrenic nerves, and the thymus gland, as well as lymph nodes and fatty and loose areolar connective tissue, and, on occasion, a portion of the thyroid and some of the parathyroid glands.

By convention, the mediastinum is divided into four portions, because of the disparate contents of these regions and the corresponding medical disorders, often tumors, that develop in them. The *superior mediastinum* is that subdivision of the mediastinum bordered by the thoracic inlet superiorly down to the level of a line drawn from the sternal-manubrial joint ("sternal angle") to the fourth intervertebral space. This contains the upper horns of the thymus, the lower lobes of the thyroid, the inferior parathyroid glands, their blood supply, and the recurrent laryngeal nerves. It also contains the top of the aortic arch, the innominate artery and vein, and the left carotid and left subclavian arteries, as well as the superior vena cava and much of the upper third of the esophagus. Masses arising in this area can be substernal goiters, thymomas, parathyroid adenomas, teratomas, thymic cysts, or germ cell tumors, as well as primary (lymphoma, Hodgkin's disease) and metastatic neoplasms of the lymph nodes of this area. By convention, the *inferior mediastinum,* that portion below the superior mediastinum, is subdivided into three portions: the *middle mediastinum,* which is defined as the contents of the pericardial sac; the *anterior mediastinum,* which lies anterior to the pericardium; and the *posterior mediastinum,* which lies posterior to the pericardium. The anterior mediastinum contains the thoracic horns of the thymus, the adjacent thymic fat pads, the phrenic neurovascular bundles, and lymph nodes. The masses most commonly arising in this region are the same as those seen in the superior mediastinum, and for this reason the anterior mediastinum is sometimes classified with the superior mediastinum as the *anterosuperior mediastinum.* The middle mediastinum contains the pericardium and the heart, and masses arising in this area are usually pericardial cysts or tumors (often metastatic) and cardiac myomas. The posterior mediastinum contains the esophagus, the thoracic duct, the thoracic sympathetic chain (nerves and ganglia), and the intercostal nerve roots, as well as the aorta and the inferior vena cava. Masses arising in this area can be bronchiogenic or esophageal cysts, rests, duplications, or diverticula, as well as esophageal neoplasms and neurogenic tumors. Thoracic aortic aneurysms can also present as a posterior mediastinal mass, and metastatic carcinomas can produce enlargements of lymph nodes in this area.

The anatomy of the mediastinum accounts for the clinical presentation of pathologic conditions arising within it. Because the mediastinum itself and its contents are invisible externally and are surrounded by bony structures, lesions arising within are often asymptomatic until they have advanced sufficiently to cause a pressure effect. Consequently, many mediastinal tumors are noted first, in asymptomatic patients, on chest radiographs (or, nowadays, CT or MRI scans), often taken for unrelated reasons. However, the rigid confines of the mediastinum cause many moderately advanced disease processes to present with symptoms caused by compression of the esophagus (dysphagia), the trachea (shortness of breath, wheezing), the recurrent laryngeal nerves (hoarseness), the sympa-

Plate 41.
Mediastinum

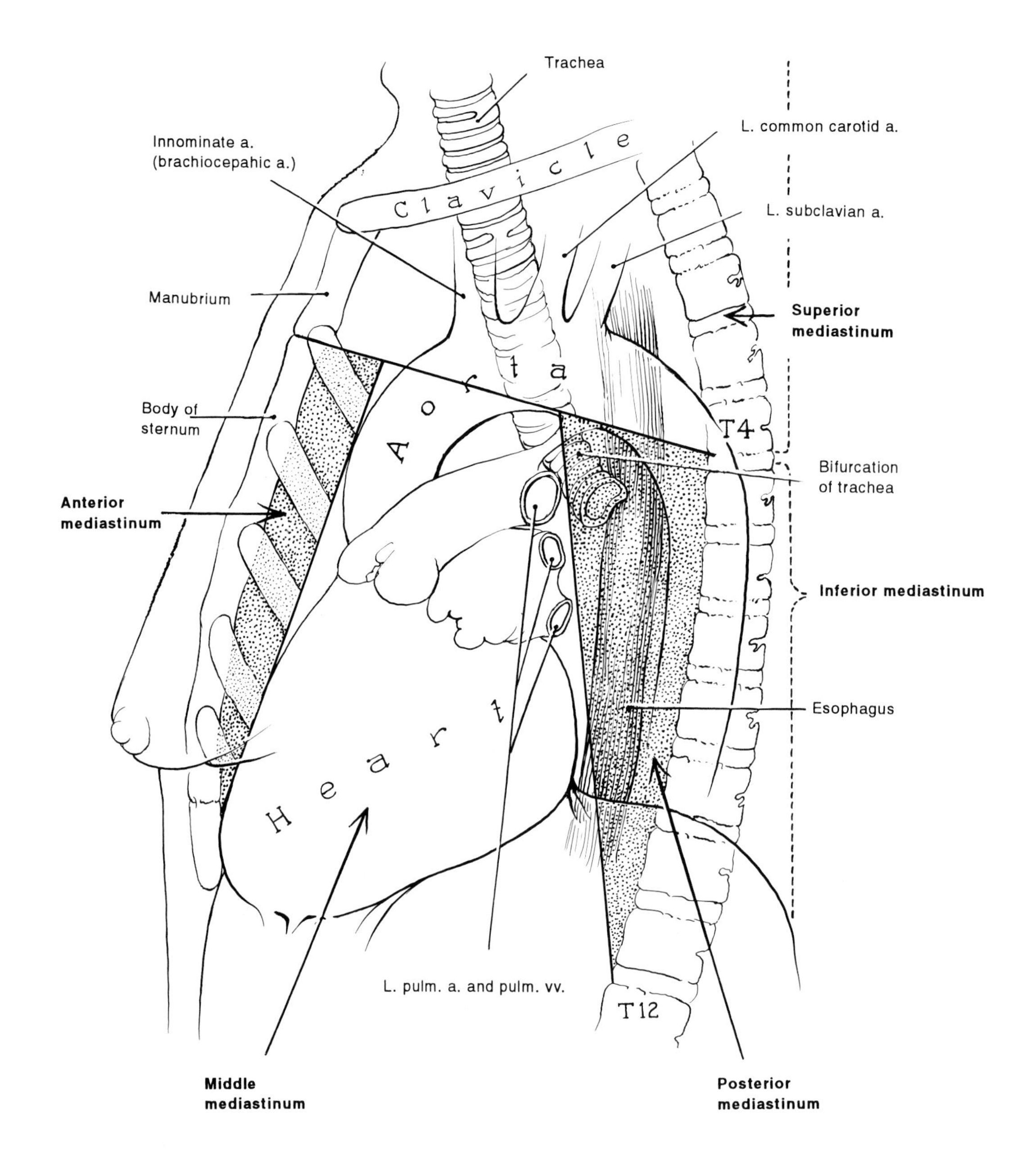

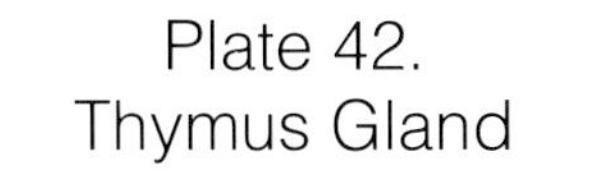

Plate 42.
Thymus Gland

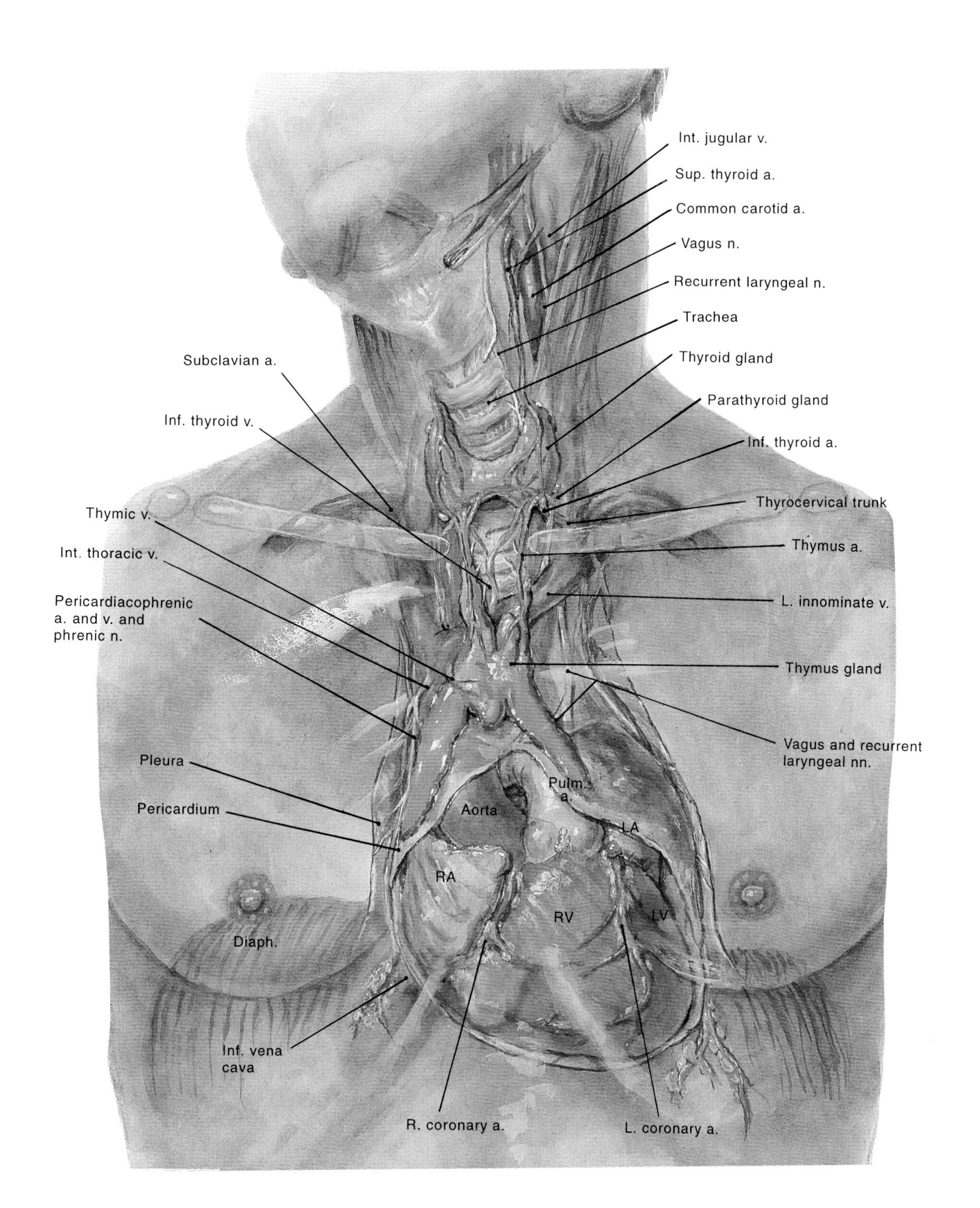

THE ANATOMICAL REGIONS OF THE MEDIASTINUM

ANATOMICAL AREA	CONTENTS	MASSES (TUMORS)
Superior mediastinum Inferior mediastinum Anterior mediastinum	great vessels, vagus, phrenic and recurrent laryngeal nerves, thymus, lymph nodes (thyroid, parathyroid)	thymomas, lymphomas, germ cell tumors, goiter, parathyroid adenomas, metastatic carcinomas
Middle mediastinum	heart, pericardium	pericardial cysts, cardiac myomas, metastatic carcinomas
Posterior mediastinum	esophagus	esophageal carcinoma
	aorta	esophageal and bronchial cysts
	inferior vena cava	duplications, rests, and diverticula
	thoracic duct, lymph nodes	aortic aneurysms
	thoracic intercostal nerve roots	metastatic carcinoma, neural tumors

thetic nervous system (Horner's syndrome), the thoracic duct (ascites), or the vena cava (superior vena cava obstruction syndrome).

The treatment of many diseases of the mediastinum, especially tumors, is largely surgical, but radiation therapy and, to a lesser extent, chemotherapy also play a role. Surgical access to the anterosuperior mediastinum may be obtained by a suprasternal incision and pretracheal, substernal dissection, with exploration and biopsy sometimes facilitated by a *mediastinoscope* and/or an elevating sternal retractor. Limited access (for biopsy) can also be obtained by *mediastinotomy,* usually via a small parasternal incision in the second or third interspace, with resection of a portion of the costal cartilage. In most cases, however, thorough anterior mediastinal exploration and access for surgery can be obtained safely only via a *median sternotomy,* whereby the sternum is divided in the middle by a specially designed, small, powered bone saw and retracted, and is then wired rigidly back together at the conclusion of the operation. This approach not only facilitates exploration and excision but also is by far the best way to avoid injury to important vascular and nervous structures (innominate artery and vein; phrenic and recurrent laryngeal nerves). In most cases a median sternotomy wound heals completely, with excellent cosmetic and full functional recovery, and, because the wired sternum is rigid immediately, there is strikingly less discomfort and ventilatory embarrassment than occurs with conventional thoracic or upper abdominal incisions. Occasionally, however, if a deep wound infection forms in the incision, purulent *mediastinitis* develops. Because of the rigid confines of the infected space, this often requires reopening and extensive bony debridement of the wound under general anesthesia, and usually reconstructive surgery employing vascularized muscle flaps. Failure to adequately control such mediastinitis can be fatal.

Surgical access to the posterior mediastinum can be by endoscopy (esophagoscopy) and thoracoscopy, but most often is via a right or left thoracotomy. In recent years, esophageal tumors have been successfully resected by a "transhiatal approach," by dissecting down from the neck and up from the diaphragm via the esophageal hiatus, effecting complete esophagectomy (with esophagogastric reconstruction by anastomosis in the neck) without sternotomy or thoracotomy. This not only has greatly reduced the operative risk in the patients with these tumors, who are frequently at high risk, but also places the anastomosis in the neck, where the complications due to a small leak are relatively minor compared to the not infrequently lethal posterior mediastinitis that can be caused by even a small leak of an esophagogastric anastomosis performed conventionally within the thorax.

THYMUS GLAND

The thymus gland, located in the anterosuperior mediastinum, arises embryologically as paired evaginations of the third branchial clefts, with the connection to the foregut obliterated soon thereafter. The resulting two thymic lobes descend substernally into the superior and anterior mediastinum and grossly (albeit not microscopically) fuse to form a single gland connected by a ventral isthmus. (This organ's descent and fusion are similar to those of the thyroid gland.) The resulting H-shaped gland consists of right and left lobes, connected at the isthmus, which is usually located directly beneath the sternum just anterior to the innominate vein. The right and left superior horns may extend, usually as thin projections directly adjacent to and associated with the inferior thyroid vessels, to and even occasionally far superior to the lower poles of the thyroid gland. (Infrequently, one or more inferior parathyroid glands may be found here, only after an exhaustive search in the conventional sites has failed to reveal the parathyroid adenoma that necessitated surgery.) The inferior, or thoracic, horns of the thymus are usually much larger than the cervical (i.e., the H-shaped thymus is asymmetrically "bottom heavy") and extend along the phrenic neurovascular bundles bilaterally, on either side of the pericardium and intimately involved laterally with the (medial) mediastinal pleura to the diaphragm. This anatomy is quite variable, and careful anatomical studies have shown that both the cervical and thoracic extensions (horns) of the thymus may be discontinuous, resulting in frequent ectopic sites of thymic tissue along this axis from the superior pole of the thyroid to the epiphrenic fat pads inferiorly.

The thymus gland is relatively large at birth but regresses in size as the child grows to adulthood. In most people this process of atrophy continues as they age, so that by the age of sixty, only a few wispy remnants of thymus tissue remain, interspersed with fat and loose fibrous connective tissue.

The major known function of the thymus gland is immunologic: it is the ultimate source of all the body's T-lymphocytes, which play a critical role in the defenses against bacterial, protozoal, viral, and fungal infections and appear to provide immune surveillance prophylaxis against the development of malignant tumors. The importance of this role is illustrated by the devastating effect that the absence of these T-cells produces when they are killed by an infection with the human immunodeficiency virus (HIV), producing the acquired immune deficiency syndrome (AIDS), an ultimately lethal disease characterized by repetitive and severe infections of all types, including those caused by common and normally nonpathogenic organisms such as yeast and *pneumocystis,* and by the development of multiple common and unusual malignant tumors, such as *Kaposi sarcoma.* Children born without a thymus gland (i.e., with *thymic aplasia*) manifest a very similar and devastating immune deficiency and usually succumb to infection at an early age, despite the best medical treatment, including the use of the most powerful antibiotics. Interestingly, the importance of the thymus for the generation of these essential T-lymphocytes decreases rapidly after birth, as the thymic lymphocyte germ cells migrate to and take up permanent residence within the bone marrow. Consequently, the surgical removal of the thymus gland after about the age of two produces no measurable or clinically recognized effect either on the immune system, on the resistance to infection, or on the development of tumors.

The one exception to this is seen in patients with *myasthenia gravis,* a not-uncommon autoimmune disease in which the body makes an antibody to the neuromuscular junction, which functionally connects the body's nerves to the muscles. This results in severe, sometimes even fatal, muscle weakness. Most patients with myasthenia gravis also have hyperplasia of the thymus gland, and 10 percent have thymic tumors *(thymomas).* More than 90 percent of these patients, often young adults, respond dramatically, although often not completely, to total removal of the thymus gland via an *extended cervicomediastinal thymectomy,* a procedure that carefully removes all of the soft tissue of the entire anterosuperior mediastinum via a median sternotomy so as to include all of the sites containing thymic tissue. Surprisingly, these patients respond equally well whether or not they have thymic hyperplasia or even a thymoma. It is not known why this operation improves myasthenia gravis, but it is clear that it does *not* work by producing broad-spectrum (non-antigen-specific) immunosuppression, because these patients do not evidence any later signs of immune deficiency (increased incidence of infection or tumors), nor do their other autoimmune diseases, such as thyroiditis, lupus, and arthritis, which are often associated with myasthenia gravis, show any signs of improvement. Therefore, the immunosuppression produced by thymectomy after about age two is apparently antigen specific (i.e., specific to the neuromuscular junction). This may be related to some *myoid cells* (cells with some of the morphologic characteristics of muscle cells) that can be found within the thymus glands of myasthenia gravis patients and of normal people, but the pathophysiologic mechanism remains unknown.

Many suspect that the thymus may well perform many other functions that are as yet not appreciated. This is because patients with tumors of the thymus can sometimes manifest a variety of paraneoplastic syndromes, producing a variety of systemic manifestations, in addition to myasthenia gravis. The fact that none of these normal thymic functions is essential, at least after age two, is indicated by the fact that, as noted above, total thymectomy patients appear to be otherwise normal in every way. Clearly, we still have a great deal to learn about this remarkable gland.

Anatomy as Viewed Laparoscopically

18

Mark A. Talamini, M.D.

The surgeon's view of anatomy is undergoing a dramatic change. In the past, general surgeons visualized anatomy directly with their eyes in three dimensions, often moving tissues with their hands to create different views and perceptions regarding intra-abdominal anatomy and pathology. Starting around 1990, an increasing proportion of intra-abdominal operations began to use minimally invasive technology. This alters the surgeon's view: the eyes no longer view the tissue directly; instead, image-capturing electronic equipment is interposed between the tissue and the surgeon's eyes, and the surgeon views the anatomy on a screen. This change has important implications. Most operations carried out in this manner deprive the surgeon of normal three-dimensional relationships. The surgeon must determine spatial relationships of tissues using clues other than three-dimensional vision. Fine detail, however, is actually improved in many instances. The image is magnified about eighteen times, so the surgeon can actually see tissues and structures better. In open abdominal surgery, the view of the intra-abdominal organs is determined by the position and size of the incision. In laparoscopic procedures, however, the view is dictated by the position of the laparoscope within the abdominal cavity. Color data are also filtered through the eye of the video telescope. Not only might colors viewed in this manner be different from colors interpreted directly by the eye, but they might also differ from case to case and vary according to the equipment used.

These changes do not affect just visualization. Most minimally invasive surgery is accomplished by creating a pneumoperitoneum. That is, the potential space within the peritoneal cavity is inflated with a gas, usually carbon dioxide, to create an area in which to see and work. The forces necessary to create this space can alter the anatomy. For instance, the anterior abdominal wall is stretched outward, which turns something that is normally a relatively flat structure into a spherical structure. The pressure on the diaphragm elevates it significantly toward the head, creating an abnormally elevated arch. The pressure on the outer bowel wall tends to collapse the bowel.

Using current technology, the surgeon cannot view the entire intra-abdominal cavity in one visual field at one time. The view is dictated by the angle of the telescope, which can display only one quadrant or region of the intra-abdominal space. This chapter is divided into sections discussing, respectively, laparoscopic anatomy viewing the abdomen in the direction of the groins, laparoscopic anatomy viewing the upper abdominal region of the stomach and esophagus, and laparoscopic anatomy viewing the region of the gallbladder. These three views were chosen as regions that are commonly observed and operated on using minimally invasive technology in its current phase of development.

LOWER ABDOMINAL ANATOMY

Minimally invasive operations focusing on the lower abdomen include laparoscopic hernia repair, laparoscopic rectal surgery, laparoscopic bladder surgery, laparoscopic lymph node removal, and laparoscopic appendectomy. Of these, laparoscopic hernia repair has generated particular interest.

A hernia is a defect in the abdominal wall. There are a number of natural sites of weakness where hernias may occur. The umbilicus, where the umbilical cord passed through the abdominal wall at birth, is one such site. In men, the path followed by the cord structures through the internal and external inguinal rings represents an area of weakness, since the testicle traverses this path-

Plate 43.
Internal Inguinal Ring

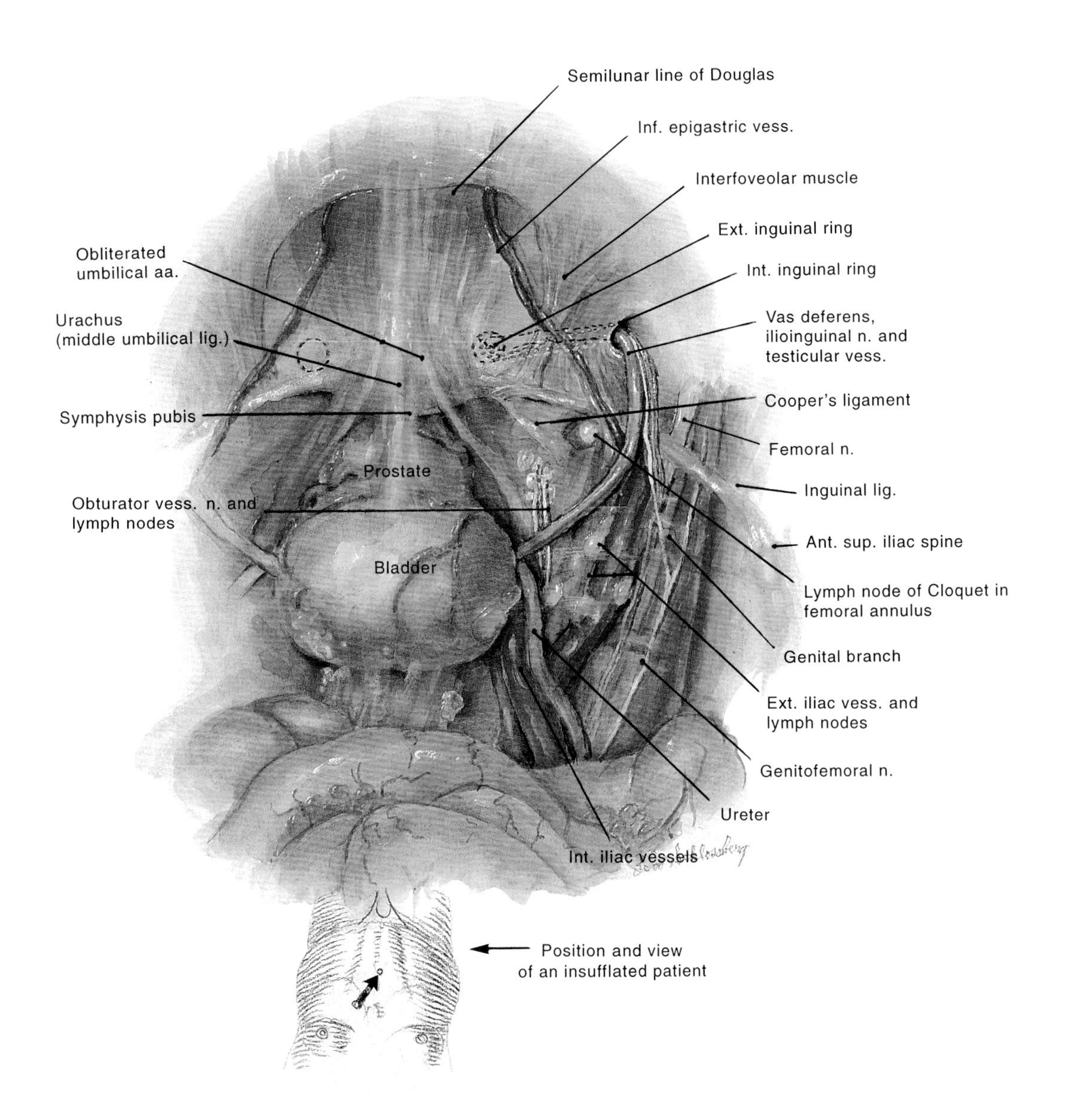

way during embryonic development. This potential weak spot often develops into a hernia defect later in life. When this tunnel remains patent at birth, infant and child hernias result.

Laparoscopic procedures such as laparoscopic hernia repair have caused surgeons to learn a new approach to the anatomy of the lower abdomen. Upon placement of the telescope into the abdominal cavity at the umbilicus, when viewing the lower abdomen, the surgeon immediately sees omentum draped over small or large bowel to a greater or lesser degree. To facilitate the view of the pelvis laparoscopically, where hernia problems and other types of pathology exist the bowel must be moved away from this area. This is usually accomplished by placing the patient in a rather steep foot-up, head-down position (the Trendelenburg position) so that gravity pulls the omental apron and the contents of the small bowel up toward the head, revealing the pelvic structures as displayed in Plate 43. A visual feature that often serves as an orienting landmark is the middle umbilical ligament, or urachus. In the majority of adults, this structure is merely a fibrous band running from the umbilicus to the superior middle portion of the bladder. It is the remnant attachment of the allantois from embryonic life. The symphysis pubis is also an important landmark for the laparoscopic surgeon. This structure can be not only visualized but also palpated with a long laparoscopic instrument. It can also be palpated from outside the abdomen with a finger, the indentation thus created being visible via the laparoscope inside the abdomen. The symphysis pubis is the structure formed by the joining of the two superior rami of the pubic bones in the midline. It is a central anchoring point in the lower abdomen, with the powerful rectus abdominus and external abdominal oblique muscles being attached there. The internal inguinal ring is the site of an indirect inguinal hernia. Here the vas deferens, the ileoinguinal nerve, and the testicular vessels join to pass through the abdominal wall musculature in a diagonal direction, extending medially and inferiorly toward the scrotum. A small circular opening in the wall allows the passage of these structures out through the inguinal canal toward the scrotum in the male. This was indeed the route followed by these structures from the abdominal cavity out to the scrotum during embryonic life. The internal inguinal ring is the entrance to this canal. The external inguinal ring is the site of exit of these structures from the inguinal canal into the scrotum.

The internal ring, as visualized laparoscopically from within the abdomen, is an important focus of attention for the laparoscopic surgeon undertaking to repair an inguinal hernia. The vas deferens, ilioinguinal nerve, and testicular vessels converge at the internal ring to form the spermatic cord. During dissection, the surgeon must be careful to avoid damage to these structures, particularly the vas deferens, which carries sperm from the testicle up through the external and internal rings, down to the seminal vesicles and the prostate gland, and finally out to the urethra. Injury to the vas deferens can have an impact on future fertility.

The inferior epigastric blood vessels are also an important anatomical feature. Each inferior epigastric artery is a direct branch from the right or left external iliac artery, a pair of arteries that are the primary continuation of the abdominal aorta as it branches. Each external iliac artery supplies blood to the lower limb and to the anterior abdominal wall. Each internal iliac artery supplies its side of the pelvis and the perineum. The external iliac artery passes directly under the inguinal ligament on its journey toward the leg. Once it passes under the ligament, it is known as the femoral artery. During laparoscopic surgery it is not directly visible under the peritoneal membrane, except in very thin patients. The surgeon must know the anatomical locations of these vascular structures so they can be avoided. The inferior epigastric artery emerges from the external iliac artery close to the inguinal ligament and can be seen arising along the inner aspect of the peritoneum toward the umbilicus. It lies just medial to the internal ring and therefore is a key landmark indicating the location of the internal ring. As the artery travels superiorly, it passes through the posterior fascia and supplies branches to the powerful rectus abdominus muscles. The surgeon classifies hernias according to their location with respect to the inferior epigastric artery. Hernia defects lateral to this blood vessel are indirect inguinal hernias (those that travel the same path as the spermatic cord). Those medial to this blood vessel are direct inguinal hernias. The external iliac artery and vein are accompanied by numerous lymph nodes, which can be particularly important in patients with prostate cancer. During some laparoscopic operations, these lymph nodes are carefully harvested to determine whether cancer has involved the lymph nodes.

The bladder occupies the center of the lower region of the laparoscopic view of the lower abdomen. It is most easily identified by the presence of the bladder catheter balloon placed just before surgery to keep the bladder decompressed. The lateral and upper margins of the bladder are often not clearly visible, so the surgeon, who must take care to avoid injury to these structures, must understand where the margins of the bladder normally lie. Below the visible surface of the bladder lies the prostate. It is accessible laparoscopically only by dissecting the upper and anterior aspect of the bladder away from its preperitoneal attachments. Deep in the abdomen and toward the midline, the left and right obturator vessels and lymph nodes are also visible. These are often the first branches of the internal iliac arteries, as they exit the pelvis near the edge of the bladder. They supply the posterior aspect of the upper thigh and the hip joint. *Cooper's ligament* is primarily a surgical term referring to the *pectineal* region of the pelvis bone as it spreads laterally from the symphysis pubis. This structure is hard, and holds stitches and staples well. Thus, it is an important structure for anchoring hernia repairs. During laparoscopic procedures, it can be identified visually with some dissection. Its nature and location can be confirmed by palpating it with a laparoscopic instrument while pressing on the symphysis externally to confirm the position of the landmark. Above Cooper's ligament, arching from medial to lateral, is the semilunar line of Douglas. This also is a surgical term referring to the ligamentous portion of the transverse abdominal muscle as it inserts onto the symphysis pubis. It is another strong structure frequently used in the repair of inguinal hernias.

The nerve structures running through the region of the lower pelvis are not visible laparoscopically. The surgeon must simply be familiar with their normal course to avoid injury during operations in this region. The femoral nerve is perhaps the most reliable, since it travels along with and lateral to the femoral artery and vein. The genitofemoral nerve travels along the psoas muscle under the inguinal ligament medial to the femoral nerve and eventually down to the lower limb. The genital branch dives deep into the pelvis from the main trunk of the genitofemoral nerve before its passage under the inguinal ligament. The ureter traverses the retroperitoneum and joins the bladder near its lateral superior aspect. It should be well away from any hernia surgery, but in thin patients it is often visible through the peritoneum as it crosses the psoas muscle and the ileac artery and vein and finally joins the bladder.

The anterior-superior iliac spine is the lateral anterior crest of the pelvic bone. It is an important and easily identified external anatomical landmark. During laparoscopic surgery, the surgeon will often poke the tissue immediately adjacent to this structure and correlate its position from within to gain important information.

The lower abdomen and groin region is a busy area, with numerous structures of importance. Arteries, nerves, veins, and elements of the urinary tract all converge in this region, which is dominated by the massive pelvis bone. In this region, anatomy is not merely an intellectual exercise. Successful surgery in this region depends on the accuracy of the surgeon's knowledge of the locations of these structures as they traverse the surfaces of the pelvis.

THE ESOPHAGEAL HIATUS, SPLEEN, STOMACH, AND VAGI AS SEEN LAPAROSCOPICALLY

As surgeons have become comfortable with laparoscopic surgery, they have undertaken more complex operations using laparoscopic technology. A prime area of interest and activity for these procedures is the upper abdomen. Operations on the spleen, the stomach, the esophagus, and the nerves regulating the production of acid in the stomach are all feasible and are all currently being pursued using laparoscopic instrumentation. Accurate identification of the anatomical structures is essential to successful surgery. Laparoscopic technology offers the advantage of a magnified view of the anatomy. This helps overcome the disadvantages of two-dimensional vision and the lost sense of direct touch and tissue manipulation.

The laparoscopic view of this region of the abdomen is different from the view during open surgery. When the upper abdomen is viewed with the laparoscope, the umbilicus is the most common fulcrum point for the device. Therefore, the surgeon's view is usually looking up from below. The initial dominant feature of the anatomical landscape seen in this region via the laparoscope is the liver. The left lobe of the liver overlies the esophageal hiatus and the fundus of the stomach and often reaches over to nearly touch the upper pole of the spleen. Therefore, if those structures are to be visualized it must be retracted out of the way with a laparoscopic instrument. When retracted, it reveals tissue surrounding the esophageal hiatus, the lesser curve of the stomach, the celiac axis arterial trunk of the aorta and its branches, and the upper regions of the stomach.

The diaphragm muscle forms the ceiling of the abdomen. It is a broad, flat muscle shaped somewhat like a bifid Chinese hand fan, divided into two large segments for the right and left sides of the abdomen. The broad, wide edges of the fanlike muscle insert circumferentially around the inner aspects of the lower rib cage. The muscle fibers converge at the handle of the fan and arch posteriorly to insert directly onto the spine. The right diaphragm insertion is referred to as the right crus of the diaphragm. It can be seen laparoscopically under the left lobe of the liver just to the patient's right side of the esophagus. The left crus of the diaphragm arches around to the patient's left and inserts on the spine as well. A gap between these two bands of muscle allows for the passage of the esophagus from the thorax into the abdomen. This gap is referred to as the esophageal hiatus. Dissection between the esophagus and the right crus of the diaphragm is the usual starting point for operations involving the esophagus or upper stomach. To aid in identification during laparoscopic surgery, a large mercury-filled rubber tube (Maloney dilator) is often passed down through the esophagus. When this tube is moved slowly in and out during surgery, the laparoscopic surgeon can see this motion, which permits precise identification and location of the esophagus. As the tissue between the esophagus and the crus of the diaphragm is dissected, the vagus nerves become visible. The left vagus nerve travels on the anterior surface of the esophagus as it passes through the esophageal hiatus and then travels down onto the anterior wall of the stomach, branching every

Plate 44.
Esophageal Hiatus

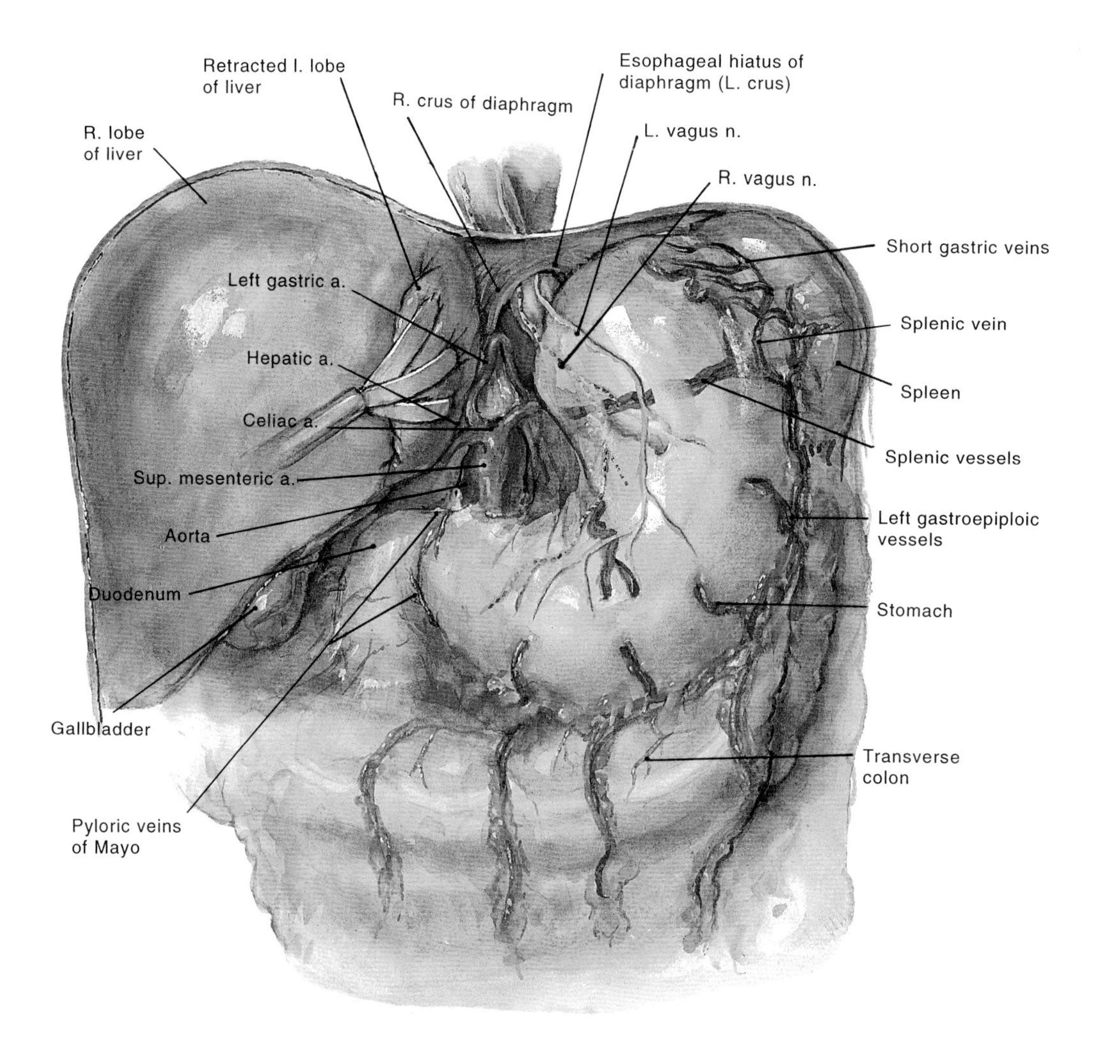

few centimeters toward the left of the stomach. The right vagus nerve traverses the esophageal hiatus nearly posterior to the esophagus. It may be located anywhere in the tissues between the posterior esophagus and the aorta. It traverses the posterior wall of the stomach, also branching repeatedly to the patient's left before passing on down the gastrointestinal tract. The vagus nerves play a role in the production of acid in the stomach, and a variety of operations have been devised to divide them at various locations to reduce acid production.

The spleen is visible as a curved purple organ tucked well into the left upper quadrant adjacent to the left diaphragm. It can be best seen with a side-viewing laparoscope up over the corner of the greater curve of the fundus of the stomach. It is intimately attached to the upper greater curve of the stomach via a web of tissue containing the short gastric veins, which drain blood from the stomach toward the spleen, eventually joining the splenic vein to return to the liver via the portal vein. The short gastric veins are fragile, broad, and easily ruptured if tissues in the region are tugged on too aggressively. These veins must be divided surgically if the spleen is to be removed laparoscopically or if the fundus of the stomach is to be used for operations to treat severe acid reflux and heartburn. The spleen itself carries a large volume of fast-moving blood. The primary exit for this blood is the splenic vein. It leaves the spleen traveling behind the pancreas and passes posterior to the stomach, eventually joining the portal vein carrying blood to the liver. The arterial blood supply to the spleen is the splenic artery, a branch of the celiac axis. It travels from the celiac axis to the patient's left along the superior edge of the pancreas. The left gastroepiploic artery supplies blood to the greater curve of the stomach. This is a main trunk blood vessel that feeds branches directly into the stomach surface at periodic intervals. The border between the stomach and the duodenum is marked by a regulating muscle valve called the pylorus. During open surgery, this muscle can be palpated with ease. In laparoscopic surgery, it is more clearly identified by visualizing the pyloric vein of Mayo—a large vein that rings the pylorus externally. The gut tube beyond this vein is the duodenum itself. The gallbladder is visualized peeping out from under the right lobe of the liver. The transverse colon is the large midportion of the colon that divides the abdomen into upper and lower regions. It is intimately attached to the greater curve of the stomach by the lesser omentum. The left gastroepiploic blood vessels actually pass through this filmy tissue structure.

The aorta passes deep to all of these structures. It is directly beneath the esophagus and can be palpated laparoscopically during the dissection between the right crus of the diaphragm and the esophagus. As it travels down through the abdomen, its first major branch is the celiac axis, which further branches into the hepatic artery, the splenic artery, and the left gastric artery. The left gastric artery supplies blood to the upper lesser curvature of the stomach. The hepatic artery supplies arterial blood to the left and right lobes of the liver as well as the gallbladder. The second major branch of the aorta is the superior mesenteric artery. It passes deep to the stomach, the duodenum, and the transverse colon, branching into multiple tributaries to supply most of the arterial blood to the small bowel and the proximal portion of the colon.

THE GALLBLADDER AS SEEN LAPAROSCOPICALLY

Removal of the gallbladder (laparoscopic cholecystectomy) is the most common laparoscopic procedure performed by general surgeons. In general, it is a very safe operation with a quick recovery. However, misidentification of key anatomical features can lead to surgical complications of tremendous significance to patients. Understanding and identification of the anatomy of the gallbladder as viewed laparoscopically are vital to safe performance of laparoscopic cholecystectomy.

Normally the anatomy of the gallbladder region is viewed from a telescope inserted at the umbilicus. Thus, the angle of view is from the umbilicus toward the right upper quadrant. When the telescope is first inserted, a view magnified approximately eighteen times is initially observed. The liver is the most dominant anatomical feature. The surgeon may or may not see the robin's-egg blue of a normal-appearing gallbladder peeping out from under the edge of the liver. Once appropriate additional cannulas are placed, the fundus of the gallbladder is grasped and retracted into the right upper quadrant. This also retracts the right lobe of the liver in an upward direction, since these two organs are intimately connected. Another retractor grasps the infundibulum region of the gallbladder and pulls it away from the portal structures, thus revealing the region of the portal triad.

Unfortunately, although the occasional patient may be thin enough to reveal all of the structures exactly as illustrated here, in many patients all of these structures are covered with an opaque layer of fat, making immediate identification impossible. When faced with this problem the surgeon must dissect the fat away carefully until the appropriate structures are revealed.

Stored bile travels from the gallbladder through the cystic duct, which enters the common bile duct in an end-to-side fashion. (Above the junction with the cystic duct, this duct is referred to as the common hepatic duct, while below this junction it is referred to as the common bile duct.) The common bile duct itself is the primary conduit of bile from the liver to the duodenum, where bile aids in the digestion of fat.

The liver is divided into the left and right lobes. As bile is collected from each lobe of the liver, it drains via the

Plate 45.
Gallbladder

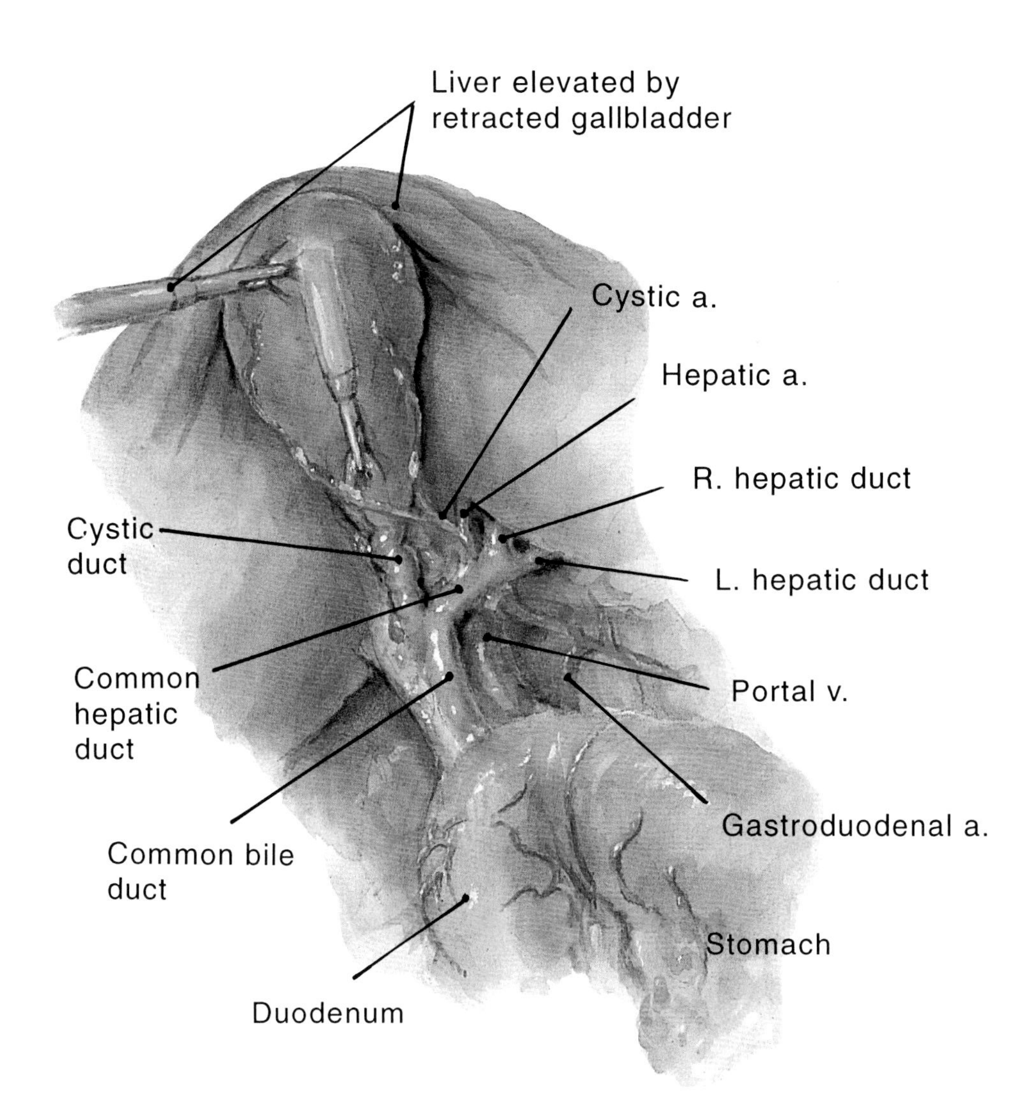

left or right hepatic duct. These two structures join to form the common hepatic duct. As the bile drains further toward the duodenum, the cystic duct joins this drainage tree in an end-to-side fashion. The most important structure that must be carefully identified is the cystic duct of the gallbladder. This is the duct that is isolated, ligated with clips, and divided as the first step of removal of the gallbladder. It is identified through careful dissection along the obvious lower surface of the gallbladder downward until the junction between the cystic duct and the gallbladder is clearly identified. The cystic duct is the primary conduit of bile between the gallbladder and the common bile duct. The duodenum is the end destination of the common bile duct. (The duodenum is the first portion of the small bowel; it is joined directly to the outlet of the stomach.)

The blood supply of the liver also passes through this key region. The portal vein carries nutrient-enriched blood from the gut directly to the liver, where nutrients are extracted for processing. It is a large blood vessel with delicate thin walls. It lies deep to the common bile duct (closer to the patient's back) and travels in a parallel direction. The liver also receives an oxygen-rich arterial supply of blood from the aorta via the celiac axis. One branch of the celiac axis, the gastroduodenal artery, supplies blood to the duodenum and a portion of the pancreas. Another branch, the hepatic artery, supplies arterial blood to the liver. Like the biliary drainage system, this artery branches into right and left hepatic arteries to supply each lobe of the liver. The main blood supply to the gallbladder is the cystic artery, a branch of the right hepatic artery. This arterial branch must be clipped and divided carefully adjacent to the gallbladder during removal of the gallbladder.

The gallbladder is enveloped entirely by a peritoneal lining. About 50 percent of the surface of the gallbladder is intimately attached to the underside of the liver. After the cystic duct and the cystic artery have been divided, the gallbladder must be carefully dissected away from the underside of the right lobe of the liver. This involves dividing this peritoneal lining investment around the circumference of the gallbladder, as well as dividing multiple tiny blood vessels communicating directly between the liver bed and the gallbladder. This is usually accomplished with an electrocautery instrument that quickly and effectively seals these vessels with minimal bleeding.

The Circulatory System

Robert K. Brawley, M.D.

The circulatory system consists of the heart, a pump (described in chapter 21), and the blood vessels, a network of tubes that carry blood to and from the heart (Plate 46). Blood vessels are designated arteries, arterioles, capillaries, or veins, depending upon their size, composition, and function. Arteries vary in size from large, named vessels to very small and unnamed vessels, but all carry blood from the heart to the body tissues. The walls of the arteries contain three layers: adventitia (outer layer), media, and intima (inner layer). The media of the larger arteries (aorta, innominate, subclavian) contains large amounts of elastic tissue and few smooth muscle cells, while the media of the medium-sized arteries (radial, popliteal, superior mesenteric) and the smaller, unnamed arteries have many smooth muscle cells and relatively little elastic tissue. Smooth muscle cells are particularly prominent in the walls of the arterioles, which are extremely small vessels about 0.2 mm in diameter and just visible without magnification.

The smooth muscle cells of the arteries and the arterioles are innervated by autonomic nerves. Impulses from the sympathetic nerves cause the smooth muscle to contract, thus diminishing the vessel diameter, while impulses from parasympathetic nerves produce smooth muscle relaxation and result in dilatation of the vessels; blood flow to various regions of the body is regulated in this way. The coronary arteries are exceptions to this rule, since in these vessels sympathetic impulses cause dilatation and parasympathetic impulses produce constriction. Arterioles lead into capillaries, vessels approximately 1 mm long that have a lumen about the diameter of a red blood cell (7 to 8 microns). The capillary wall consists of a single layer of endothelial cells, which permits exchange of oxygen, carbon dioxide, nutrients, and waste products between the body tissues and the blood. Veins carry blood from the capillaries to the heart. Usually, veins have the same name as the adjacent artery. Although vein walls have the same three layers as arteries, the media of the veins is poorly developed, and thus the walls of veins are less thick than the walls of arteries of similar diameter.

The aorta is the main trunk of the systemic arterial system, and its parts are designated the ascending aorta, the arch of the aorta, and the thoracic and abdominal portions of the descending aorta (Plate 47). The ascending aorta, located in the middle mediastinum, begins at the aortic valve and ends at the innominate artery (brachiocephalic trunk). Its only branches are the left and right coronary arteries. The arch of the aorta lies in the superior mediastinum and has three branches: the innominate artery, the left common carotid artery, and the left subclavian artery. After a short distance, the innominate artery divides into the right subclavian and right common carotid arteries. The arteries that provide the major portion of blood supply to the head and neck are the right and left common carotid arteries, each of which divides into two branches: the external carotid, supplying the neck, the face, and the exterior of the head; and the internal carotid artery, supplying the anterior brain, eye, orbit, and sinuses. The subclavian artery brings blood to the upper extremities. It becomes the axillary artery at the lower border of the first rib, and at the lower border of the axilla it becomes, in turn, the brachial artery, which divides into the radial and ulnar arteries at the elbow. Branches of the subclavian artery include the vertebral and the internal mammary arteries (internal thoracic artery).

The descending thoracic aorta traverses the posterior mediastinum and gives rise to intercostal, bronchial, esophageal, pericardial, mediastinal, and diaphragmatic branches. The intercostal arteries course below each rib and form anastomoses with the intercostal branches of the internal mammary arteries. Blood is supplied to the spinal cord principally by the anterior spinal artery but also by the two posterior spinal arteries (Plate 47). The anterior spinal artery is formed at the level of the medulla oblongata from branches of the two vertebral arteries. This artery descends along the anterior aspect of the spinal cord and gives off numerous circumferential and penetrating branches, which provide the blood supply to the spinal cord itself. The posterior spinal arteries take origin from the cerebellar arteries and descend along the posterior aspect of the spinal cord. They also send penetrating branches into the spinal cord. These three spinal arteries are joined at numerous sites along their courses by radicular spinal branches of the vertebral, cervical, intercostal, lumbar, and sacral segmental arteries (Plate 47). Of specific interest is the artery of Adamkiewicz, which is a relatively large radicular spinal branch of the lower intercostal or upper lumbar arteries. The artery of Adamkiewicz connects with the anterior spinal artery in the vicinity of the crura of the diaphragm and provides important blood supply to the spinal cord. Interruption of the posterior spinal arteries may produce no untoward effect because of extensive anastomoses. However, interruption of the anterior spinal artery or the artery of Adamkiewicz will often result in paralysis.

As the aorta passes through the aortic hiatus of the diaphragm, it is designated the abdominal portion of the descending aorta, or, more simply, the abdominal aorta. The first branches of the abdominal aorta are the phrenic arteries, which supply the inferior surface of the diaphragm. The celiac artery arises from the aorta just below the aortic hiatus and has three major branches: the common hepatic artery, the splenic artery, and the left gastric artery. As their names imply, these branches provide blood supply to the liver, the spleen, and a major portion of the stomach. In addition, much of the duodenum and pancreas receive blood from branches of these arteries. The superior mesenteric artery arises from the aorta about 1 cm below the celiac artery and supplies the entire small intestine except for the proximal duodenum; it also gives blood supply to the cecum, the ascending colon, and the proximal one-half of the transverse colon. The renal arteries take origin from the aorta just distal to the superior mesenteric arteries. The abdominal aorta also gives origin to the suprarenal, testicular, or ovarian arteries and four pairs of lumbar arteries. The lumbar arteries supply blood to the lateral and posterior abdominal walls and to the spinal cord via spinal branches. The inferior mesenteric artery takes origin from the abdominal aorta about 3 or 4 cm above the aortic bifurcation, and supplies the distal one-half of the transverse colon, the descending colon, the sigmoid colon, and a major part of the rectum. At the level of the fourth lumbar vertebra the abdominal aorta bifurcates into the left and right common iliac arteries. The middle sacral artery, a small vessel, arises from the posterior wall of the aorta just proximal to its bifurcation. The common iliac arteries divide to form the internal and external iliac arteries. The internal iliac, or hypogastric, artery supplies structures of the pelvis, the buttock, the generative organs, and the medial aspect of the thigh. The external iliac artery passes along the wall of the pelvis and at the inguinal ligament becomes the common femoral artery, which is the major source of blood supply for the lower extremity. The femoral triangle is a space formed by the inguinal ligament and the border of the sartorius and adductor longus muscles; the femoral artery, vein, and nerve course through it. The common femoral artery divides into the superficial femoral artery and the deep femoral, or profunda femoris, artery about 3 cm below the inguinal ligament. This latter artery is a major source of blood supply to the structures of the thigh. The superficial femoral artery traverses Hunter's canal in the thigh and becomes the popliteal artery at the knee. Below the knee, the popliteal artery divides into three branches: the anterior tibial artery, the posterior tibial artery, and the peroneal artery. These arteries supply the lower leg and the foot.

The superior vena cava is the great vein that receives blood from the upper body and returns it to the heart; the inferior vena cava serves the same function for the lower part of the body. The internal jugular veins receive blood from the brain, face, and neck, while the subclavian veins collect blood from the upper extremities. The internal jugular and the subclavian veins join to form the innominate (brachiocephalic) veins, which, in turn, join to form the superior vena cava.

The veins of both the upper and lower extremities are classified as either deep or superficial, depending upon their position relative to the deep fascia of the extremity. In the lower extremity, the superficial venous system consists of the long and short saphenous veins and their tributaries (Plate 48). The long and short saphenous veins connect with one another through several communicating veins. The deep venous system is composed of the popliteal and femoral veins and their tributaries. The superficial and deep venous systems communicate at the knee, where the short saphenous vein passes through the popliteal fascia to enter the popliteal vein, and at the groin, where the long saphenous vein joins the femoral vein by penetrating the fascia at the fossa ovalis. In addition to these two principal sites of communication, the superficial and deep systems are connected at numerous points by unnamed perforating veins that pierce the fascia of the leg. Both superficial and deep veins of the lower extremities contain numerous valves. The femoral veins receive tributary veins from the lower extremities and, as they pass under the inguinal ligaments, become the iliac veins, which also collect blood from the pelvic structures. The right and left common iliac veins join to form the inferior vena cava, which, as it traverses the abdominal cavity, is joined by the right spermatic or ovarian vein, renal veins, suprarenal veins, hepatic veins, and lumbar veins. The portal vein is formed by the merging of the superior mesenteric and splenic veins, which receive blood from intestines, pancreas, and spleen (Plate 46, insert). Blood from the portal vein passes through the liver and is collected by the hepatic veins, which join the inferior vena cava just below the diaphragm.

The function of the circulatory system is to provide blood flow that brings oxygen and nutrients to and carries carbon dioxide and waste products away from peripheral tissues. A region of the body is said to be ischemic when blood flow to that area is inadequate to allow normal function. In the presence of severe ischemia, cellular metabolism, which is usually aerobic, becomes anaerobic. Carbon dioxide and lactic acid accumulate, and metabolic acidosis develops in the region. When cardiac output (the amount of blood pumped by the heart into the arteries each minute) becomes abnormally low, the body alters the distribution of blood flow and shunts blood to more vital body areas, such as the heart and brain, and away from the musculoskeletal and splanchnic vascular beds. This shunting of blood is accomplished, in part, through the autonomic nervous system by constricting or dilating arteries, particularly arterioles, thereby changing resistance to blood flow in various body regions.

As the blood is pumped through the arteries by the heart and as it returns to the heart through the veins, it exerts a certain pressure within the blood vessels. Arterial blood pressure is a product of the cardiac output and the peripheral resistance to blood flow that exists in the arterial system at any given moment. Cardiac output is largely determined by the adequacy of the heart as a pump and by the blood volume. Resistance is primarily a function of the status of the arterioles, but it is also influenced by the diameter of the muscular arteries, the elasticity of the larger arteries, and viscosity of the blood. Hemorrhage that seriously depletes blood volume and myocardial damage that diminishes the power of the heart can result in decreased cardiac output and an abnormally low blood pressure, i.e., hypotension or shock. A variety of pathologic conditions can cause increased peripheral resistance due to severe constriction of arterioles and produce abnormally high blood pressure, i.e., hypertension.

Once the blood passes through the capillaries, most of the energy of cardiac contraction has been expended to overcome the resistance of the arterioles and capillary beds. In the supine subject, a small pressure gradient exists between the peripheral veins and the right atrium and propels blood toward the heart. In the erect position, the force of gravity acts to retard venous return to the heart from the lower extremities, and the "pumping" mechanism of the muscles of the legs and the existence of valves in the veins of the extremities become important factors in preventing stasis of blood in the venous system (Plate 48). Thrombophlebitis can destroy the valves of the veins and result in venous stasis, varicose veins, and skin ulcers. During walking the gastrocnemius and soleus muscles contract, thereby increasing pressure in the veins and lymphatics of the calf. The increased pressure tends to propel blood in the veins and tissue fluids in the lymphatics toward the heart. As the muscles of the leg relax, pressure in the veins diminishes and blood in the veins begins to flow toward the feet. This retrograde blood flow is prevented by the venous valves, which close and hold the blood in that segment of vein until the next muscular contraction again forces blood antegrade toward the heart. Soldiers standing at attention for long periods of time are known to be susceptible to fainting due to pooling of blood in their lower extremities. They are usually advised to tighten their calf muscles periodically to help move blood from legs to the heart.

Plate 46.
Composite Anatomy of the Vascular System, Based upon Arteriograms and Venograms

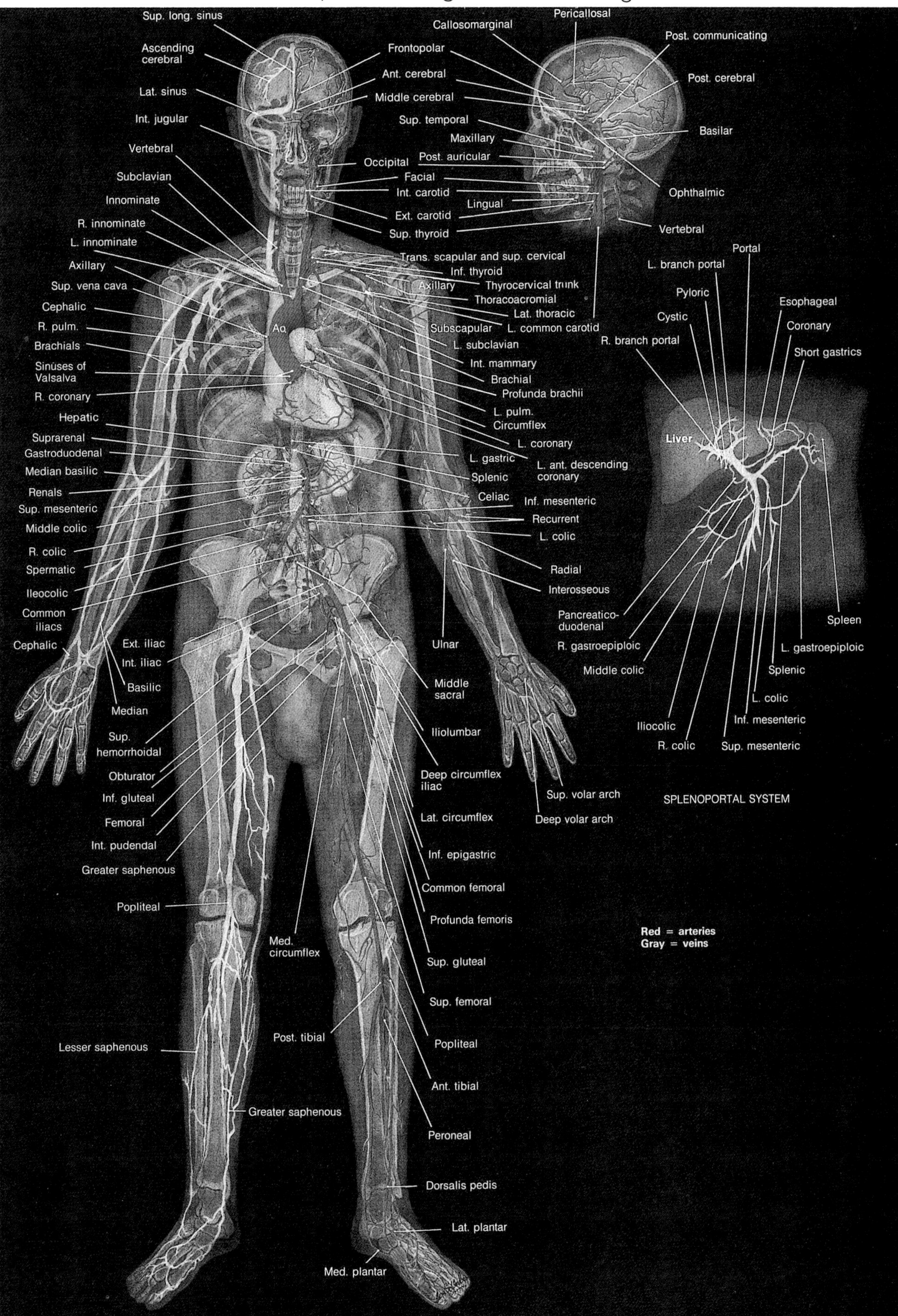

Plate 47.
The Aorta and Branches–Anterior View

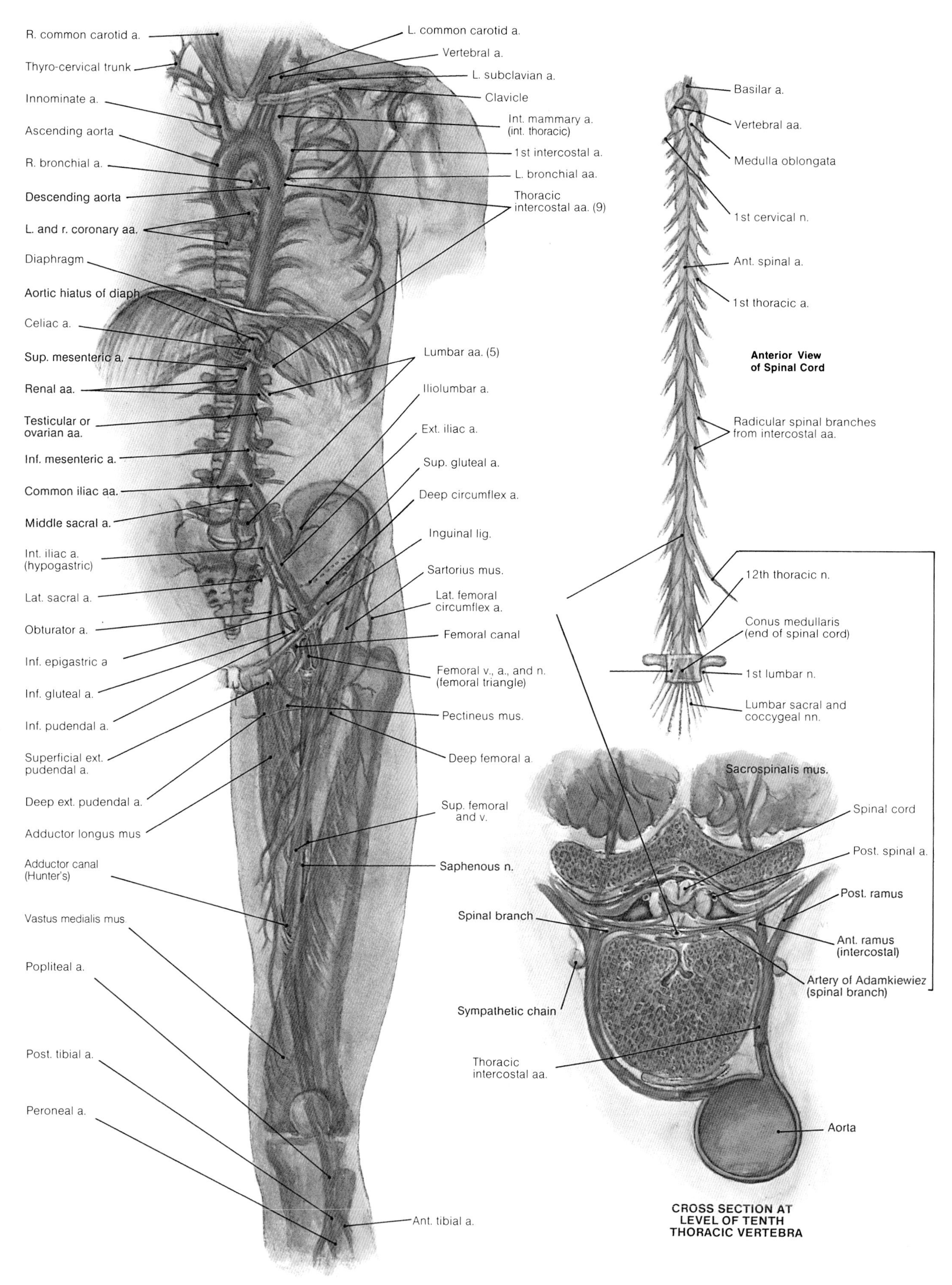

Plate 48.
Muscular Venous Pump of the Leg

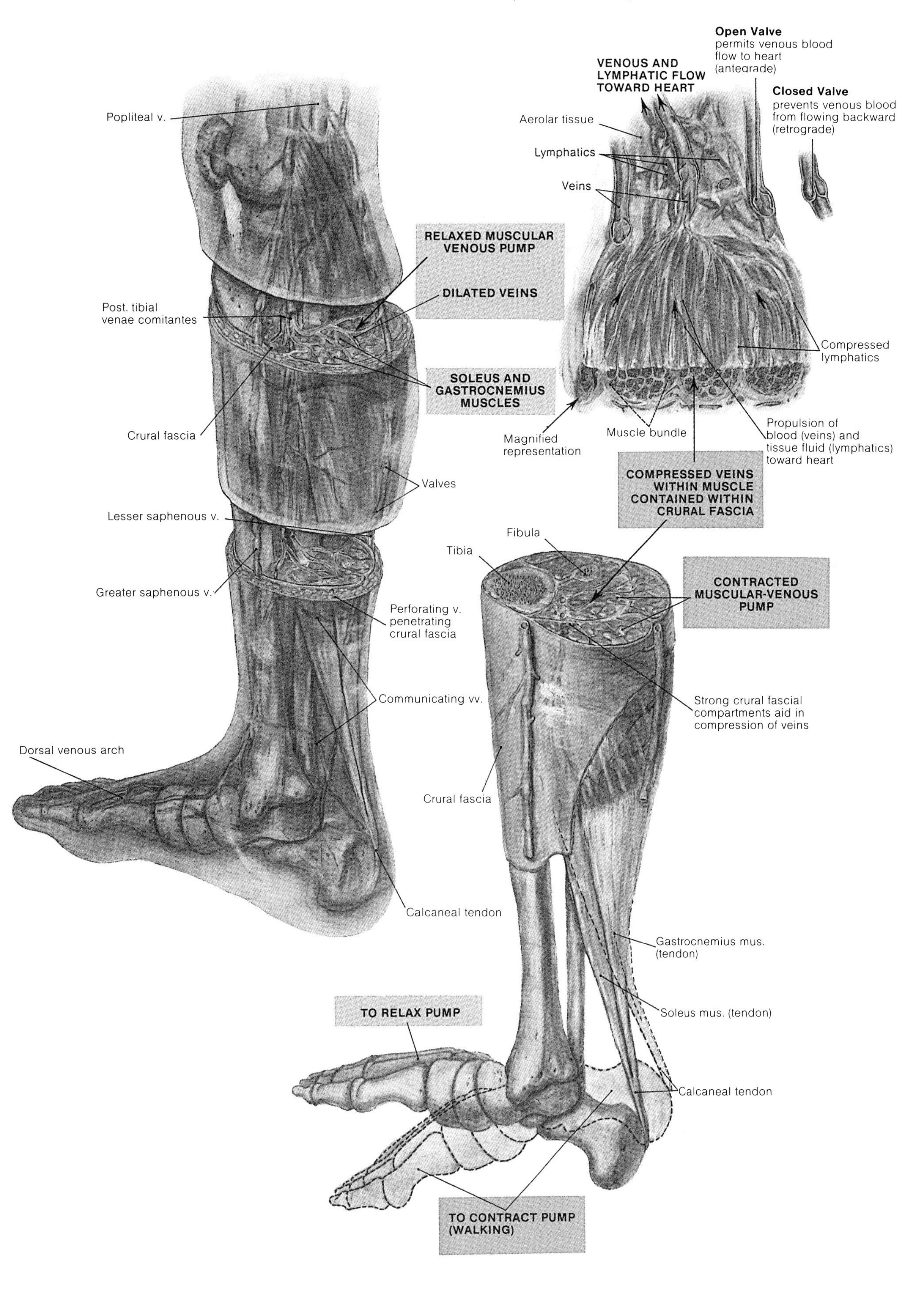

The Breast

R. Robinson Baker, M.D.

In the young nulliparous female, the breast sits as a conical protuberance on the chest wall extending in the midclavicular line between the second rib superiorly and the seventh rib inferiorly (Plate 49). The bulk of the breast tissue is in the upper outer quadrant of the breast, and a tongue of breast tissue from the upper outer quadrant extends through a defect in the fascia into the lower axilla. It should be noted that the protuberance is composed of the mammary gland enclosed in a fatty mass. More important, the mammary gland extends beyond the protuberance of the breast superiorly, medially, laterally, and inferiorly.

The predominant external feature of the female breast is the nipple-areola complex, which is centrally located. The areola surrounds the nipple as a wrinkled, pigmented region of skin. Areolar glands (Montgomery's glands) are seen within the skin of the areola as slightly raised bumps, and their secretions serve to lubricate the nipple during nursing.

With increasing age and number of pregnancies, the breast becomes more pendulous and loses its conical appearance. Microscopically the breast consists of an epithelial collection of acini and their respective ducts. Although there is tremendous variation in size, breasts always consist of fifteen to twenty lobes of glands interspersed in a loose configuration of fat and connective tissue. The amount of fat in the breast varies, and as a woman reaches the postmenopausal years the breast becomes less glandular and more fatty. Ducts draining the individual glands of each lobe coalesce and eventually form the lactiferous duct (2 to 4 mm in diameter) located behind the nipple-areola complex. The lactiferous duct proceeds toward the nipple, and each duct ends as a separate orifice on the surface of the nipple.

The mammary gland tissue is encased within the superficial fascia and lies anterior to the deep fascia of the chest wall (Plate 50). Bands of the posterior superficial fascia fuse with the fascia of the muscles, forming the posterior suspensory ligaments of the breast.

One of the distinctive features of the glandular architecture of the breast is the presence of "toothlike" projections of fibrous connective tissue that extend outward from the deep superficial fascia on the chest wall to the anterior superficial fascia and the overlying dermis. These bands, known as Cooper's ligaments, act to give the breast form and mobility on the chest wall. Cooper's ligaments are of clinical significance because invasion by a malignant tumor will shorten them, thus pulling in the overlying skin. Retraction of the skin is pathognomonic of malignancy.

CHEST WALL AND AXILLA

It has often been stated that in considering the surgical anatomy of the breast, one must also understand the anatomy of the chest wall. The text that follows will emphasize this important relationship.

The spaces between the twelve ribs that form the bony structure of the chest are termed the *intercostal spaces;* they contain the external and internal intercostal muscles and the intercostal vessels and nerves. The external intercostal muscles can be recognized by the direction of their fibers, which are angled slightly forward. As these muscles approach the costochondral junctions of the ribs with the sternum, they form a membranous sheet.

The superficial muscles of the chest are primarily the muscles that attach the shoulder girdle to the chest wall (see Plate 51). For example, the pectoralis major muscle arises from the medial half of the clavicle, just below the deltoid muscle; from the costal margin of the sternum (from the second to the sixth ribs); and from the upper part of the aponeurosis of the external oblique

Plate 49.
Anatomy of the Breast

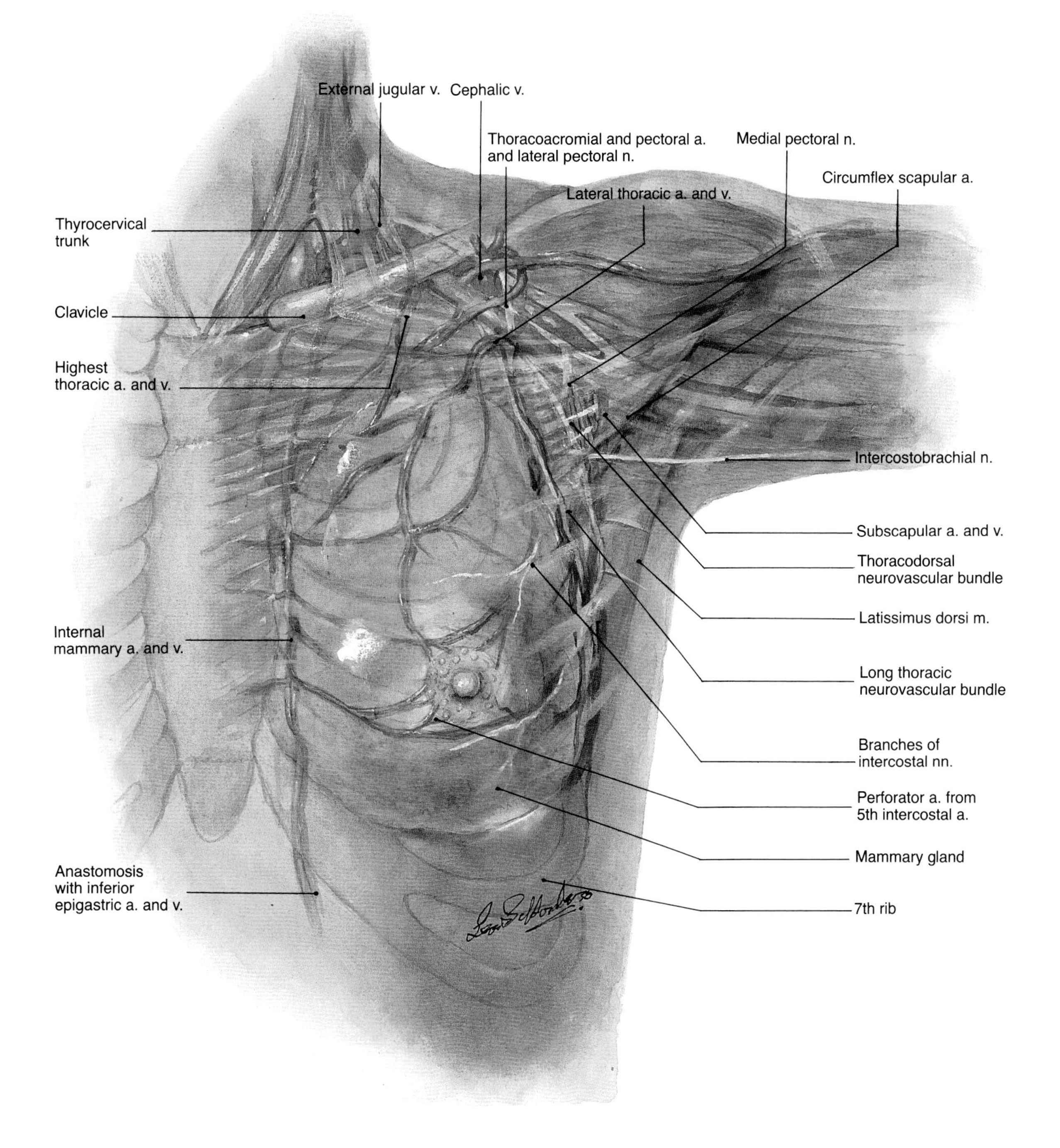

Plate 50.
Sectional Views of the Breast

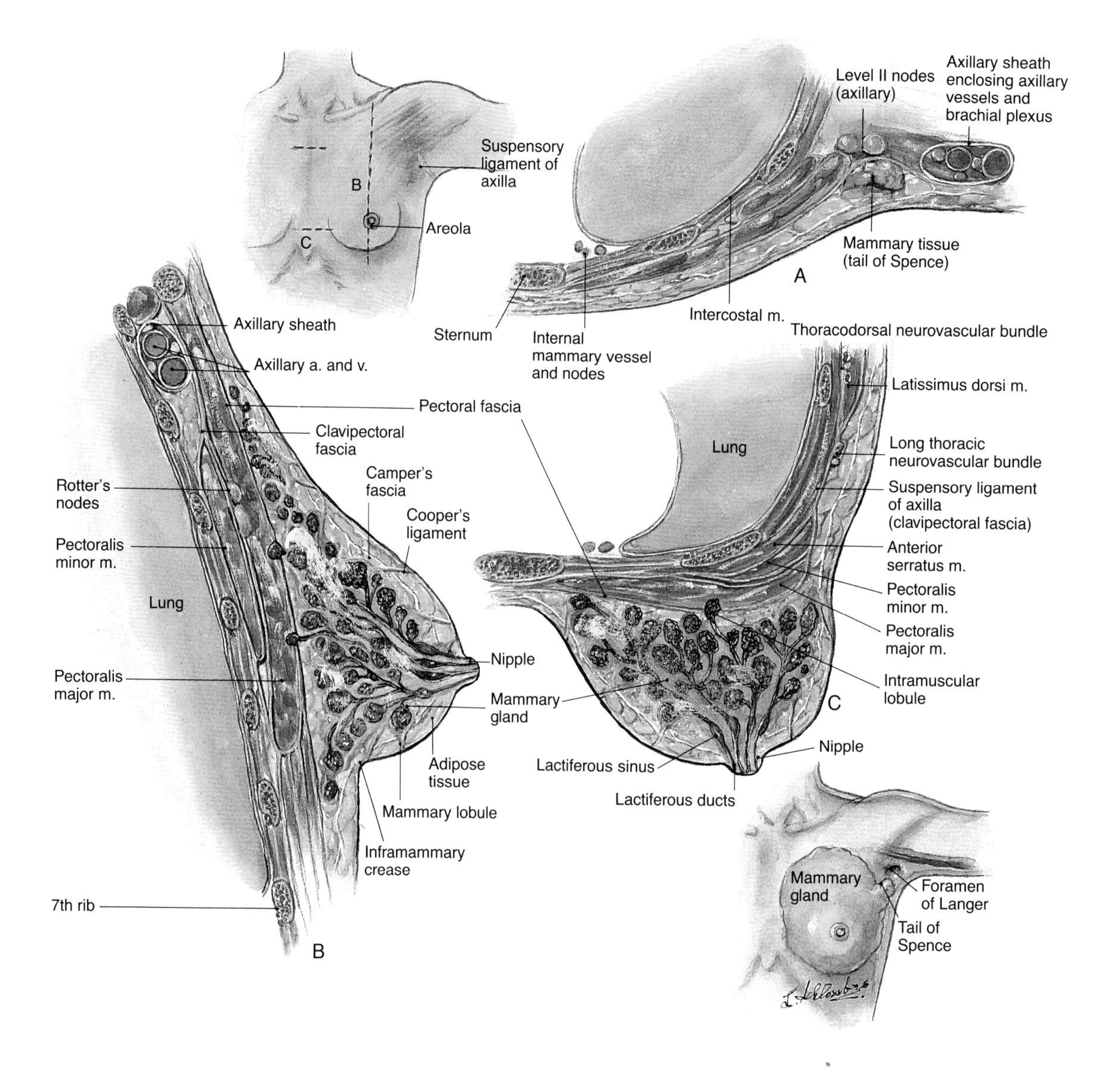

muscle. The pectoralis major muscle inserts on the upper humerus and can be divided into a clavicular head and a sternal portion. Behind the pectoralis major muscle lies the pectoralis minor muscle, which has its origin on the coracoid process of the scapula and inserts on the second through fifth ribs anteriorly. These two muscles lie just deep to the breast and form the anterior boundary of the axilla.

The serratus anterior muscle forms the medial boundary of the axilla and is composed of multiple slips of muscle originating on the anterior surface as well as the superior borders of the upper eight, nine, or sometimes even ten ribs and the fascia over the intervening intercostal spaces (see Plate 51). The sheet of muscle then wraps around the lateral chest wall to insert along the medial scapular border. The posterior boundary of the axilla is composed of the subscapularis, teres major, and latissimus dorsi muscles.

The apex of the axilla extends upward to the chest wall, where the axillary vein becomes the subclavian vein as the vessel enters the chest (at the junction of the clavicle and the first rib), and extends into the posterior triangle of the neck behind the clavicle and anterior to the scapula. The inferiormost aspect of the axilla is the skin and underlying fascia.

MAJOR ARTERIES AND VEINS

The major blood supply for the breast comes from arteries that arise from the internal mammary artery, penetrate the intercostal muscles near the sternum, and pass through the pectoralis major muscle on their way to the medial aspect of the breast (see Plate 51). It will generally be possible to identify three or four such branches. The artery entering from the second intercostal space near the sternum is almost always the largest of these branches from the internal mammary artery. The veins that accompany the arteries drain blood from the breast back to the heart through the internal mammary vein, which subsequently drains into the innominate vein.

Other arteries supplying the breast arise from branches of the axillary artery. The subclavian artery becomes the axillary artery as it passes beneath the clavicle. The axillary artery medial to the head of the pectoralis minor muscle gives rise to the supreme thoracic artery (near the clavicle); and laterally it gives rise to the thoracoacromial artery and the lateral thoracic artery. These branches arise from the axillary artery behind the pectoralis minor muscle. The thoracoacromial artery gives off a pectoral branch that descends between the pectoralis minor and major muscles on the posterior surface of the pectoralis major muscle, thus providing the arterial blood supply to this muscle. A few branches traverse the muscle and supply the deeper aspects of the breast.

The lateral thoracic artery arises either from the axillary artery or occasionally from the trunk of the thoracoacromial artery. It courses to the posterior surface of the pectoralis minor muscle by passing behind the axillary vein. It may send a branch around the lateral border of the pectoralis major muscle to the breast.

Three arteries arise from the axillary artery lateral to the lateral border of the pectoralis minor muscle: the subscapular artery and the anterior and posterior humeral circumflex arteries. The subscapular artery is the largest branch of the axillary artery, and after giving off a branch to the subscapularis muscle it courses toward the lateral chest wall and the latissimus dorsi muscle. This portion of the artery is termed the *thoracodorsal artery.* In fact, most surgeons know this major axillary artery branch by this name rather than by the name *subscapular artery,* which is the term used by anatomists.

Within this area of the axilla are numerous arterial and venous branches that cross the space between the medial surface of the latissimus dorsi muscle and the serratus muscle on the chest wall. These branches must be carefully identified and divided to isolate and preserve the accompanying thoracodorsal nerve. Careful preservation of the thoracodorsal artery and vein is also important if the latissimus dorsi muscle is to be used in reconstruction.

These descriptions of the major arterial branches of the axillary artery also serve to define the several major tributaries of the axillary vein, since these venous structures correspond to their respective arterial branches. These veins drain the lateral and deeper aspects of the breast, the pectoral muscles, and the muscles of the chest wall.

Several points regarding the anatomy of the axillary vein and its major branches are worthy of emphasis. Foremost is the marked anatomical variation observed. For example, the junction of the basilic and brachial veins to form the axillary vein may be quite variable, occurring anywhere between the teres major muscle laterally and the clavicle medially. Thus, it is not uncommon for the surgeon to be presented with what appears as a "double" axillary vein on opening the clavipectoral fascia laterally. The surgeon must recognize this variation and avoid mistakenly ligating one of these trunks while assuming it to be merely a branch of the axillary vein.

The breast and chest wall also have venous drainage laterally through posterior intercostal veins that communicate with the azygos vein through the plexus of vertebral veins.

NERVE SUPPLY TO THE BREAST AND CHEST WALL

The sensory innervation of the skin of the breast occurs by way of the lateral and anterior cutaneous branches

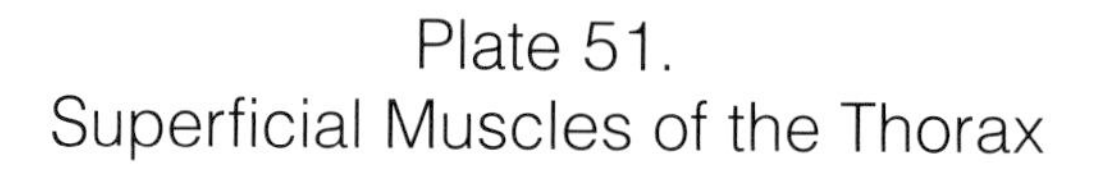
Plate 51.
Superficial Muscles of the Thorax

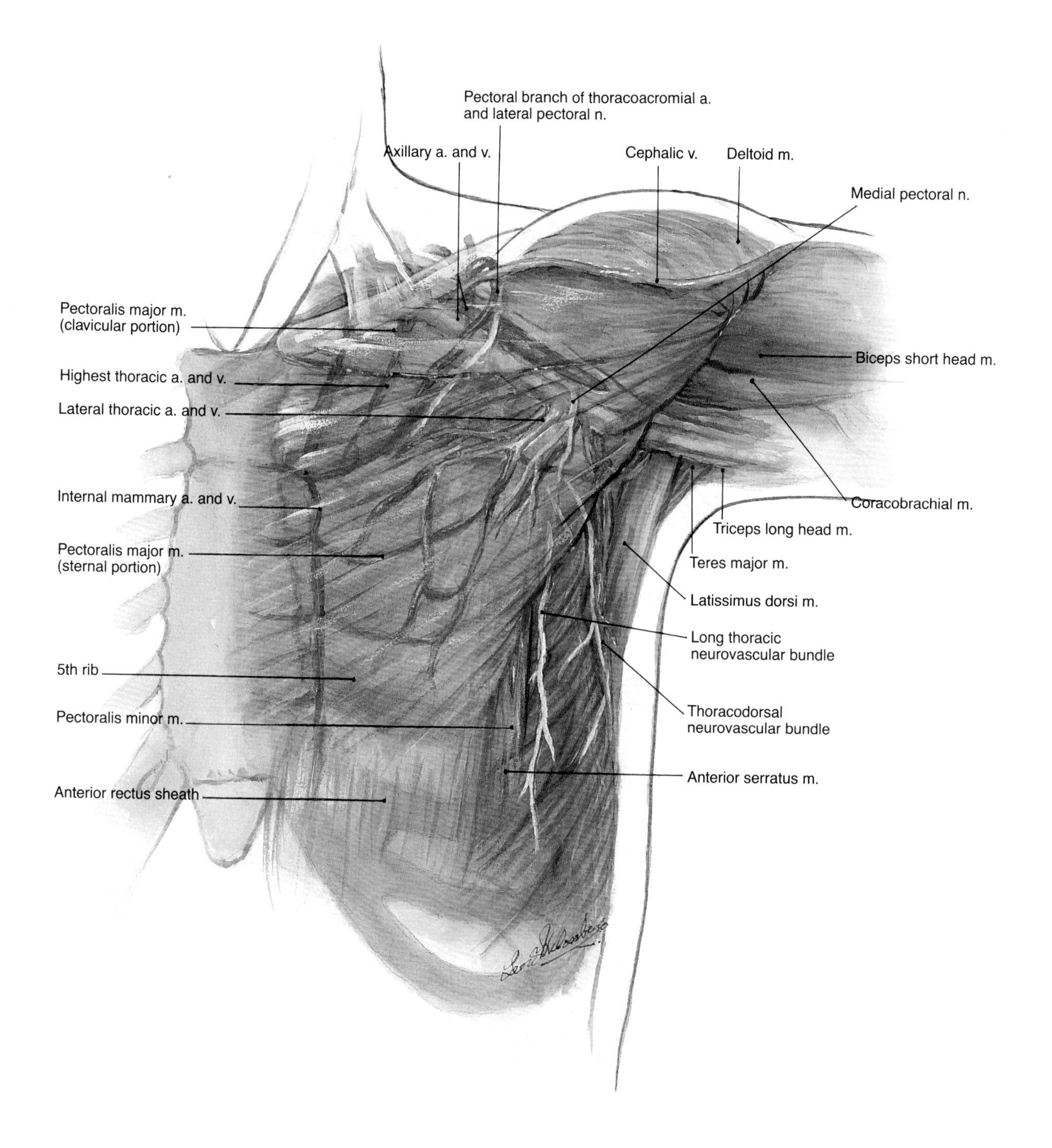

of the second through sixth intercostal nerves. Some of the skin of the uppermost portion of the breast and the infraclavicular region of the chest is supplied by branches of the supraclavicular nerve, itself a branch of the cervical plexus. The lateral portions of the breast and the lateral chest wall receive sensory innervation through the third through sixth lateral branches of the intercostal nerves. These lateral branches are near the attachment sites of the serratus anterior muscle. The anterior branches supply the skin of the medial aspect of the breast and sternal area.

The nipple-areola complex has a large number of free and Ruffini-like nerve endings and Krause's end-bulbs within the dermis. These receptors transmit tactile sensations: stretch and pressure. Stimulation of these receptors results in erection of the nipple and, in the lactating woman, a flow of milk. The latter occurs through communication of these nerves with cells in the hypothalamus. Suckling stimulates these receptors, which generate impulses that cause the release of oxytocin in the hypothalamus. Oxytocin stimulates the breast glands and causes contraction and ejection of the milk.

Special note is made of the lateral branch of the second intercostal nerve because of its surgical importance. This nerve is larger than the other lateral branches and gives off a branch identified by the surgeon as the intercostobrachial nerve. The intercostobrachial nerve supplies sensory fibers to the medial aspect of the upper arm. If it is sacrificed or injured during surgery, numbness or hyperesthesia of this area of skin result, but this is not a very significant deficit in most patients.

The intercostobrachial nerve can be seen exiting the fascia of the intercostal muscles in the second interspace behind the lateral edge of the pectoralis minor muscle. It crosses the axilla directly through the axillary fat pad containing level I lymph nodes, crosses over the anterior border of the upper portion of the latissimus dorsi muscle, and proceeds below the axillary vein into the upper arm, where it usually joins the medial cutaneous nerve of the arm. The second intercostal nerve frequently bifurcates and, at times, trifurcates as it crosses the axilla; it may also receive a branch from the third intercostal nerve.

The brachial plexus resides well out of the operative field and is usually not seen during the dissection (see Plate 51). Nevertheless, the surgeon must be keenly aware of its location in relation to the axillary artery and vein, as well as its anatomical configuration. The brachial plexus arises from the fifth cervical through the first thoracic ventral roots. These five spinal roots above the subclavian artery join to form three nerve trunks, which then divide into anterior and posterior divisions.

The three posterior divisions form the posterior cord. The anterior divisions of the upper and middle trunks form the lateral cord, and the anterior division of the lower trunk forms the medial cord. These three cords of the brachial plexus surround the axillary artery and, in fact, derive their names—posterior, lateral, and medial—from their relationship to the axillary artery. These structures lie behind and slightly above the axillary vein, which is the landmark for the upper limit of the surgeon's dissection of the axilla.

The surgeon needs to be familiar with four nerves that must be identified and carefully dissected during the course of an axillary lymph node dissection (see Plate 51). The first two nerves to be identified and preserved are the lateral pectoral nerve and the medial pectoral nerve. The lateral pectoral nerve derives its name from its origin, the lateral cord of the brachial plexus, even though its anatomical location in the axilla is medial to the medial pectoral nerve. The medial pectoral nerve is so called because it arises from the medial cord of the brachial plexus. The lateral pectoral nerve enters the surgical field anterior to the medial border of the head of the pectoralis minor muscle, where it is already beginning to branch, and proceeds to fan out on the posterior surface of the pectoralis major muscle. Injury to this nerve would result in significant atrophy of the body of the pectoralis major muscle.

The medial pectoral nerve courses across the axillary vein behind the head of the pectoralis minor muscle. Branches of the nerve supply this muscle. Of importance to the surgeon is the presence of a branch that either exits from the lateral aspect of the pectoralis minor muscle or, in most cases, appears as a gentle "loop" of nerve that curves around the lateral border of the pectoralis minor muscle and enters the lateral border of the pectoralis major muscle. Stimulation of this branch with a nerve stimulator will demonstrate contraction of the muscle bundles of the lateral border of the pectoralis major muscle. A characteristic of this branch as it is dissected from surrounding axillary tissue is the presence of a small artery and vein, which can be found coming off the nerve branch at the apex of the loop. Injury to this branch of the medial pectoral nerve will result in atrophy of the lateral aspect of the pectoralis major muscle. Preservation of these nerves has taken on greater significance with the increasing request for reconstruction and the desire to preserve muscle function and bulk.

The long thoracic nerve that innervates the serratus anterior muscle arises not from the cords of the brachial plexus but from the fifth, sixth, and seventh spinal roots of the brachial plexus (see Plate 51). It then courses downward to the axilla behind the axillary artery and vein, entering the axilla through the cervicoaxillary canal and lying beneath the serratus fascia. It is identified in the floor of the axilla posterior to the intercostobrachial nerve. The nerve continues for some distance along the chest wall before branching and disappearing into the muscle bundles of the anterior serratus muscle.

Finally, the surgeon must also identify the thoracodorsal nerve, which arises from the posterior cord (see Plate 51). This nerve lies behind the axillary artery and vein; it can be found just below the axillary vein and deep to the subscapular artery and vein. The nerve crosses the subscapularis muscle, to which it sends a branch, and eventually inserts as two major branches into the medial surface of the latissimus dorsi muscle. The value of identifying the long thoracic nerve and the thoracodorsal nerve and preserving their integrity is well known to the surgeon.

LYMPH NODES AND LYMPHATICS

Lymphatics from the skin of the breast and lymphatics from the mammary lobules, including a rich plexus of subareolar lymphatics, tend to follow blood vessels but in general drain to the lymph nodes of the axilla. Lymph from the breast also drains medially to the internal mammary lymph nodes, which are located in a chain behind the costochondral junctions of the ribs with the sternum.

Although there are a number of lymph node groups classically described by anatomists, it seems more practical to discuss the lymph node regions of the breast as they are commonly defined by the surgical dissection of the axilla (Plate 52). The highest (most central) group is the level III lymph nodes, which are the axillary lymph nodes located along the medial and inferior surface of the axillary vein at its entrance into the chest. This group of six to twelve lymph nodes lies behind the pectoralis major muscle and medial to the medial border of the pectoralis minor muscle.

The level II axillary lymph nodes are those that are positioned behind the pectoralis minor muscle and below the vein. Anterior to the pectoralis minor muscle, between the pectoralis minor and pectoralis major muscles, one can usually identify one to five lymph nodes in the interpectoral group, the so-called Rotter's lymph nodes.

The level I lymph nodes lie between the lateral border of the pectoralis minor muscle and the medial side of the latissimus dorsi muscle. Sometimes the group of lymph nodes medial to the latissimus dorsi muscle and caudal on the chest wall are termed the lateral mammary lymph nodes.

Plate 52.
Lymphatics of the Breast

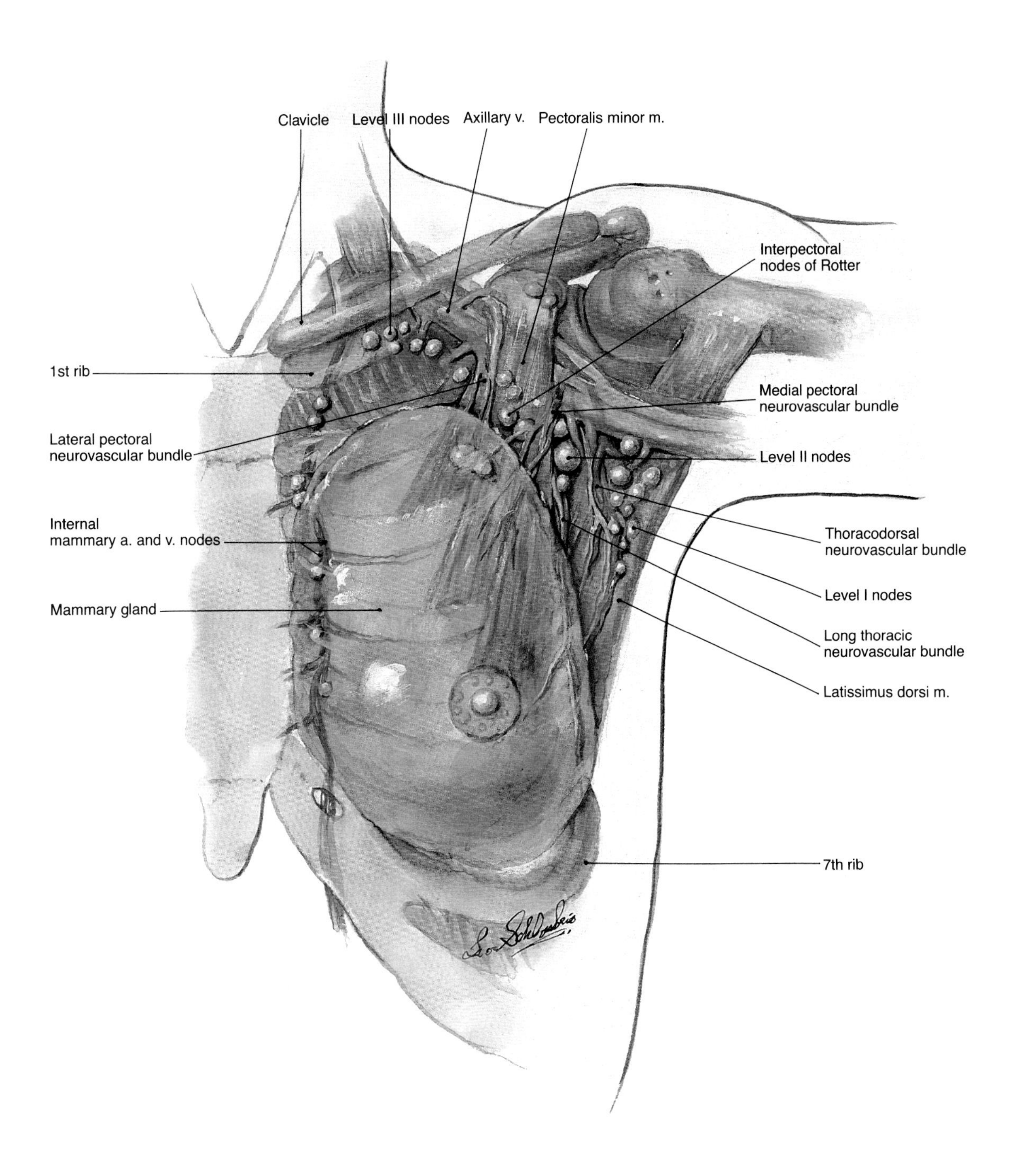

The Heart

Vincent L. Gott, M.D.

CARDIAC STRUCTURES AND FUNCTION

The adult human heart, weighing approximately 300 grams, is an extremely efficient muscular pump, designed to contract 42 million times a year and eject 700,000 gallons of blood during the same period. Although the design of this muscular pump is relatively simple from the standpoint of mechanical structure, the control mechanism for heart rate and cardiac output is exquisitely sensitive and thus quite elaborate.

The right atrium serves as the receiving chamber for all the systemic venous blood returning through the superior and inferior venae cavae. During the period of cardiac relaxation (ventricular diastole), blood flows from the right atrium into the right ventricle, and then, during ventricular systole, this blood is ejected through the pulmonary valve into the pulmonary circulation. The two valves (tricuspid valve and pulmonary valve) serve simply to permit a one-way flow of blood in an efficient manner through the right heart and into the pulmonary vasculature. In the normal heart, the pressure rises to approximately 5 mm Hg in the right atrium during contraction of this chamber. Pressure within the right ventricle is normally 25/0 (systole/diastole). Since there is no perceptible gradient across any of the normal heart valves, pressure in the pulmonary artery is normally 25/10.

Oxygenated blood returns from the lungs to the left atrium through the pulmonary veins and then, during ventricular diastole, flows through the mitral valve and into the left ventricle. Again, during ventricular systole, blood is ejected by the left ventricle into the ascending aorta through the aortic valve. The aortic valve is very similar in appearance to the pulmonary valve in that it has three delicate cusps, or pockets, which again permit blood flow in only one direction. The tricuspid and mitral valves, on the other hand, have leaflets tethered by fibrous cords (chordae tendineae) attached to the papillary muscles in the apex of each of the two ventricles. The tricuspid valve, of course, has three separate cusps, and the mitral valve has two leaflets.

Normal pressure in the left atrium during atrial contraction is approximately 15 mm Hg, and, during left ventricular contraction, the pressure in this latter chamber rises to approximately 120 mm Hg, with a similar systolic pressure in the ascending aorta. Ordinarily, the period of systole takes less than 0.3 second and the period of diastole is approximately 0.7 second. During diastole, pressure in the aorta falls to a level of approximately 80 mm Hg.

The right and left coronary arteries arise from the root of the aorta, with the orifices of each of these vessels being situated in the sinuses of the aortic valve. The left coronary artery courses behind the pulmonary artery and divides within 2 cm into the anterior descending artery and the circumflex artery. The right coronary artery courses on the surface of the heart between the right atrium and the right ventricle, and its blood supply terminates in the diaphragmatic surface of the heart. After coronary blood passes through the capillaries of the myocardium, approximately 60 percent of venous flow enters the right atrium through the large coronary sinus vein, and the remaining flow drains into the left atrium and the left and right ventricles through the Thebesian veins. Unlike in any other organ in the body, the greatest flow of blood takes place through the heart during the period of diastole. During the ejection phase of ventricular systole, coronary blood flow is greatly reduced because of compression of the coronary vessels by the cardiac muscle fibers.

All of the structures of the heart, of course, can be affected by disease or congenital deformity. Fifty thousand children are born every year with congenital malformations of the heart that include holes or defects in the atrial and ventricular septum and deformity of any of

Plate 53.
The Heart

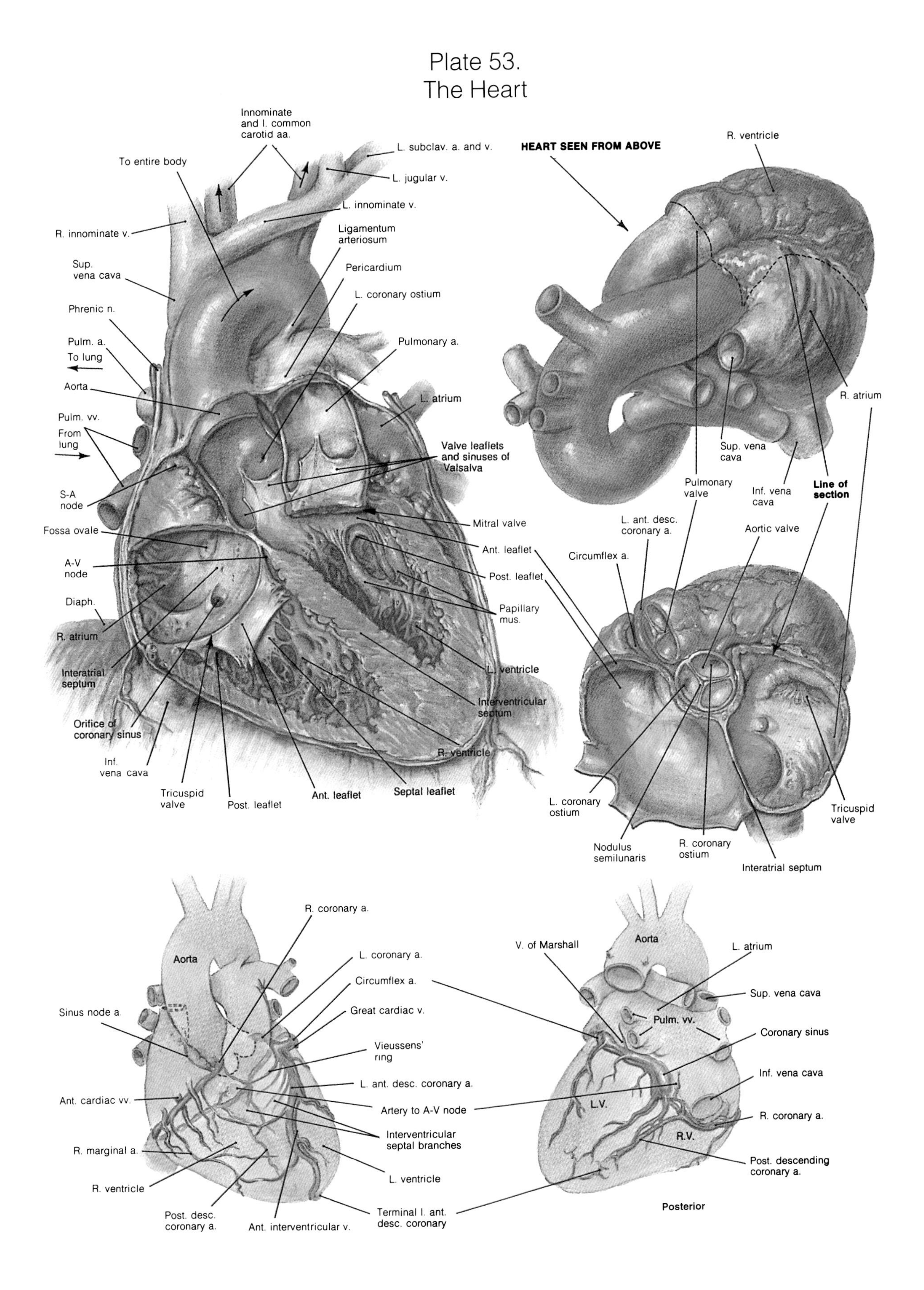

Plate 54.
Anatomy and Physiology of the Heart

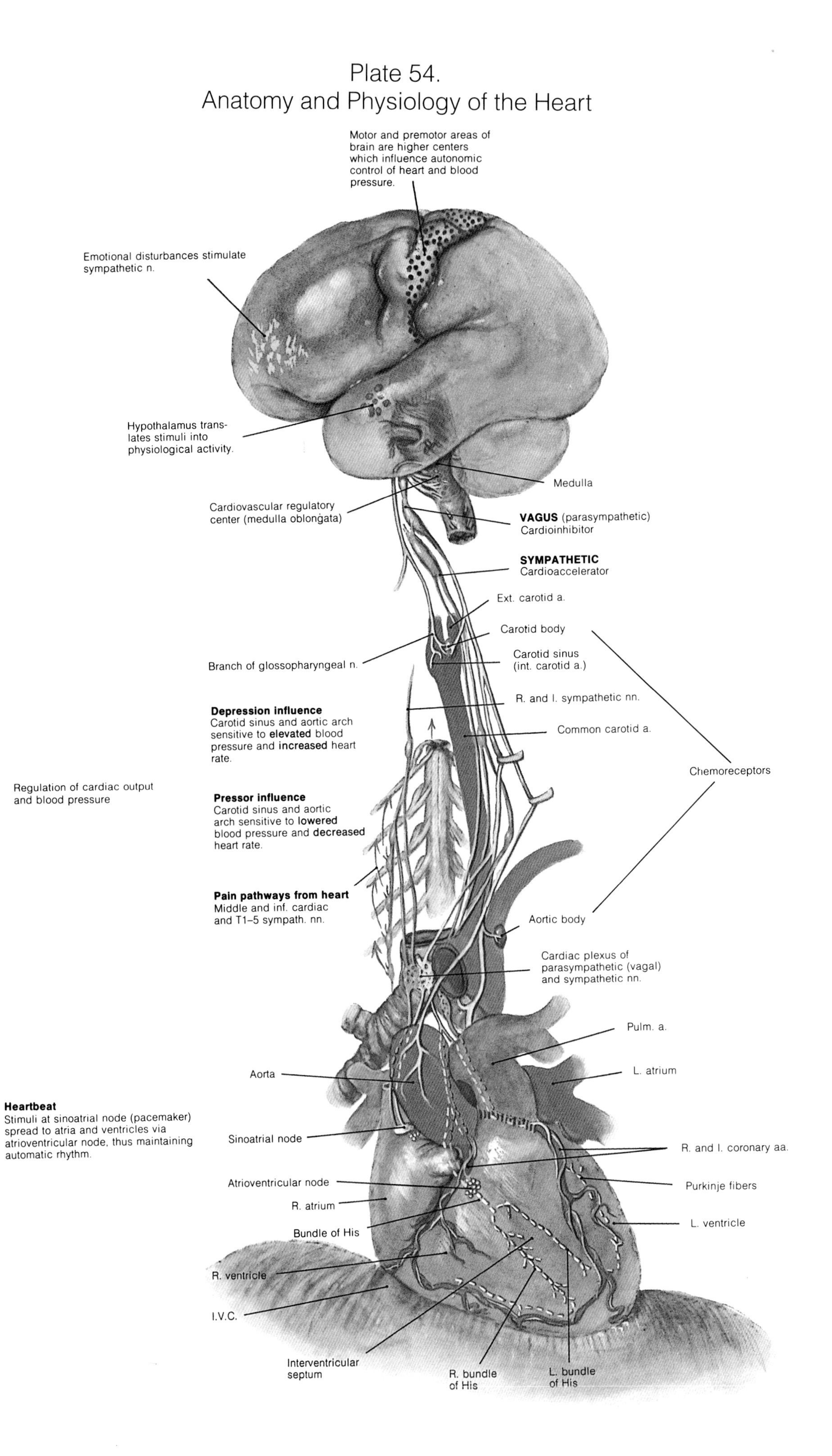

the four valves. In addition to these congenital defects, there are acquired diseases of the valves secondary to rheumatic heart disease, and, also, occlusion of the coronary arteries can result from arteriosclerosis. Most of these congenital and acquired defects can be surgically corrected, at the present time, with relatively low risk.

NEUROLOGIC CONTROL OF HEART RATE

In the mammalian heart, there is a system of specialized tissue (junctional tissue) that possesses the property of rhythmical impulse formation and conductivity to a higher degree than the cardiac muscle itself. One tiny island of junctional tissue called the *sinoatrial node* (S-A node) has this ability to generate a rhythmical electrical impulse, and it is commonly referred to as the "pacemaker" of the heart. The S-A node is located at the junction of the superior vena cava and the right atrium, and the impulse generated by this junctional tissue spreads as an electrical wave in all directions through the muscle of the atrium. This impulse is then picked up by a second node called the *atrioventricular node* (A-V node), which is located at the base of the right atrium just above the tricuspid valve. The electrical impulse then passes from the A-V node down into the A-V bundle, and into right and left branches to all portions of both ventricles through terminal Purkinje fibers. It is this excitation wave that provides the changes in electrical potential that can be recorded as the electrocardiogram.

Although the heart has a complex neuro-control system to maintain proper heart rate and cardiac output, the totally denervated heart can respond surprisingly well to changes in its load. As demonstrated in patients with heart transplants, this is achieved primarily by the effect of catecholamine release from the adrenal medulla, as well as a second important mechanism, described many years ago by the physiologist Starling. He pointed out that if the ventricles receive an increased volume of blood, they can respond by contracting more forcibly. This obviously is an important self-regulatory mechanism for the denervated heart.

In the normal innervated heart, the primary center for control of cardiac rate is the cardiovascular regulatory center located in the medulla oblongata. This center is under constant stimulation from higher centers in the brain, centers that are primarily located in the frontal lobes, where the emotional disturbances of anger, fear, and excitement are rapidly transmitted, through the hypothalamus, as an excitatory stimulus to the cardiovascular regulatory center.

The primary control of heart rate is through the vagus nerves arising in the region of the cardiovascular regulatory center. Stimulation of the vagus nerves causes an inhibitory effect with slowing of the heart rate and lowering of the blood pressure. The final vagal pathway to the heart is though the cardiac plexus, which is located at the bifurcation of the trachea. In addition to the vagal innervation (parasympathetic) of the heart, there is sympathetic innervation, which likewise arises from the cardiovascular regulatory center. These sympathetic fibers emerge from the cervical and upper thoracic ganglia of the sympathetic cord and pass by way of the superior, middle, and inferior cardiac nerves, again through the cardiac plexus, to the heart. Stimulation of the sympathetic fibers accelerates heart rate and increases the force of contraction.

There are some important afferent nervous reflexes that play a critical role in the rate of heart contraction. Afferent nerves arising in the right atrium can sense an increase in atrial pressure and stimulate the cardiovascular regulatory center to increase the heart rate (Bainbridge reflex). Similarly, there are afferent fibers, termed *pressoreceptors,* in the arch of the aorta and in the bifurcation of the carotid arteries, which sense an increased pressure in these vessels and in turn signal the cardiovascular regulatory center to slow down the heart rate (Marey's reflex). The afferent fibers from the pressoreceptors in the arch of the aorta and the carotid sinus pass via the vagus nerves (X) and glossopharyngeal nerves (IX), respectively, to the cardiovascular regulatory center. In addition, there are chemoreceptors located in the arterial wall at the carotid bifurcation and aortic arch that sense changes in blood pO_2, pCO_2, and pH. The afferent impulses arising in these fibers mainly alter respiration, but to a lesser degree modify heart rate and vasomotor tone. Decreasing pO_2 and pH and increasing pCO_2 not only increase the respiratory rate but accelerate the heart rate and enhance vasoconstriction. The afferent impulses from the chemoreceptors pass, with the afferent fibers, from pressoreceptors via the ninth and tenth nerves to the cardiovascular regulatory center.

An additional important set of afferent fibers are the pain fibers, which are stimulated by conditions of myocardial ischemia and cause the clinical condition of angina pectoris, or pain in the chest. These sensory fibers pass with the sympathetic nerves in the middle and inferior cardiac nerves and enter the spinal cord via the posterior roots of the upper five thoracic segments. This routing of afferent pain fibers of the heart through these upper five thoracic segments explains why the pain of myocardial ischemia is referred to the shoulders, upper extremities, and regions of the neck.

The Lungs

Henry N. Wagner, Jr., M.D.

In the thirteenth century ibn-an-Nafis challenged the concept of Galen that blood passes from the right side of the heart to the left via invisible pores in the cardiac septum, where it mixes with the air from the lungs to produce the vital spirit. He proposed that "in the wisdom of God," blood was carried to the lungs via the pulmonary artery so that "what seeps through the pores in the branches of this vessel into the alveoli of the lungs mixes with air and is then carried to the left chamber of the heart by the pulmonary veins." Five hundred years later, Stephen Hales elucidated the details of the relationship of structure and function in the lung. He correctly measured the dimensions of alveoli as 1/100 of an inch (254 micra) with a total surface area of about 289 square feet (27 square meters). He confirmed that this enormous surface was the site of entry of the "air particles" found in the blood and concluded that the most crucial function of the lungs is the exchange of carbon dioxide for oxygen.

Other important functions of the lung are related to the central position of the pulmonary circulation between the two chambers of the heart, and through which the entire cardiac output passes. These functions include sieving of particulate matter and the production and elimination of substances such as hormones from the blood.

The right lung is usually larger than the left, because of the space occupied by the heart. The left lung consists of two lobes, the right lung of three. The interlobar fissure of the left lung runs diagonally downward from a point about 6 cm from the apex of the lung to the base of the lung. The upper lobe consists of four segments: the apical-posterior, anterior, superior, and inferior lingular segments. The left lower lobe consists of a superior, anterior-medial basal, lateral basal, and posterior basal segment. The right upper lobe consists of an apical, posterior, and anterior segment; the middle lobe of a lateral and medial segment; and the lower lobe of a superior, anterior, medial, lateral, and posterior basal segment.

The airways, bronchial and pulmonary circulations, lymphatic system, and the nerves supplying the lungs make up the important structural elements of the lungs. In the adult, the pulmonary arteries accompany the airways quite closely, while pulmonary veins lie between two airway trees. Bronchovascular units consist of lobes, segments, lobules, acini, and alveoli. It has also been found useful to divide the lungs into three concentric zones. The respiratory zone consists of the alveoli, with the alveolar capillary network that consists of a waffle-like array of vessels with numerous connections. Here the alveoli and vessels are in intimate contact, which permits effective exchange of gases to take place. In the conductive zone are the bronchi, bronchioles, pulmonary arteries, and veins. Their walls separate air from blood, and regulatory mechanisms control the relative distributions of both, continually matching ventilation and perfusion in the healthy state. The transitory zone connects the other two and contains the respiratory bronchioles, alveolar ducts, and sacs, and the pre- and post-capillaries.

Electron microscopy has permitted measurement of the thickness of the alveolo-capillary tissue separating air from blood. In its thinnest portions it was found to be about 0.4 micra. In reaching this point, molecular gases traverse the airway system that distributes the air among the alveoli. During gaseous exchange, the tissues traversed by the respiratory gases comprise the alveolar epithelium, its basement membrane, a narrow connective tissue space, the basement membrane of the capillary endothelium, and the capillary endothelium itself. Because of the low pressure in the lungs, the slightest variation in resistance to blood flow either in the vessels themselves or transmitted to the vessels

Plate 55.
The Lungs, Bronchi, Pleurae, and Blood Vessels

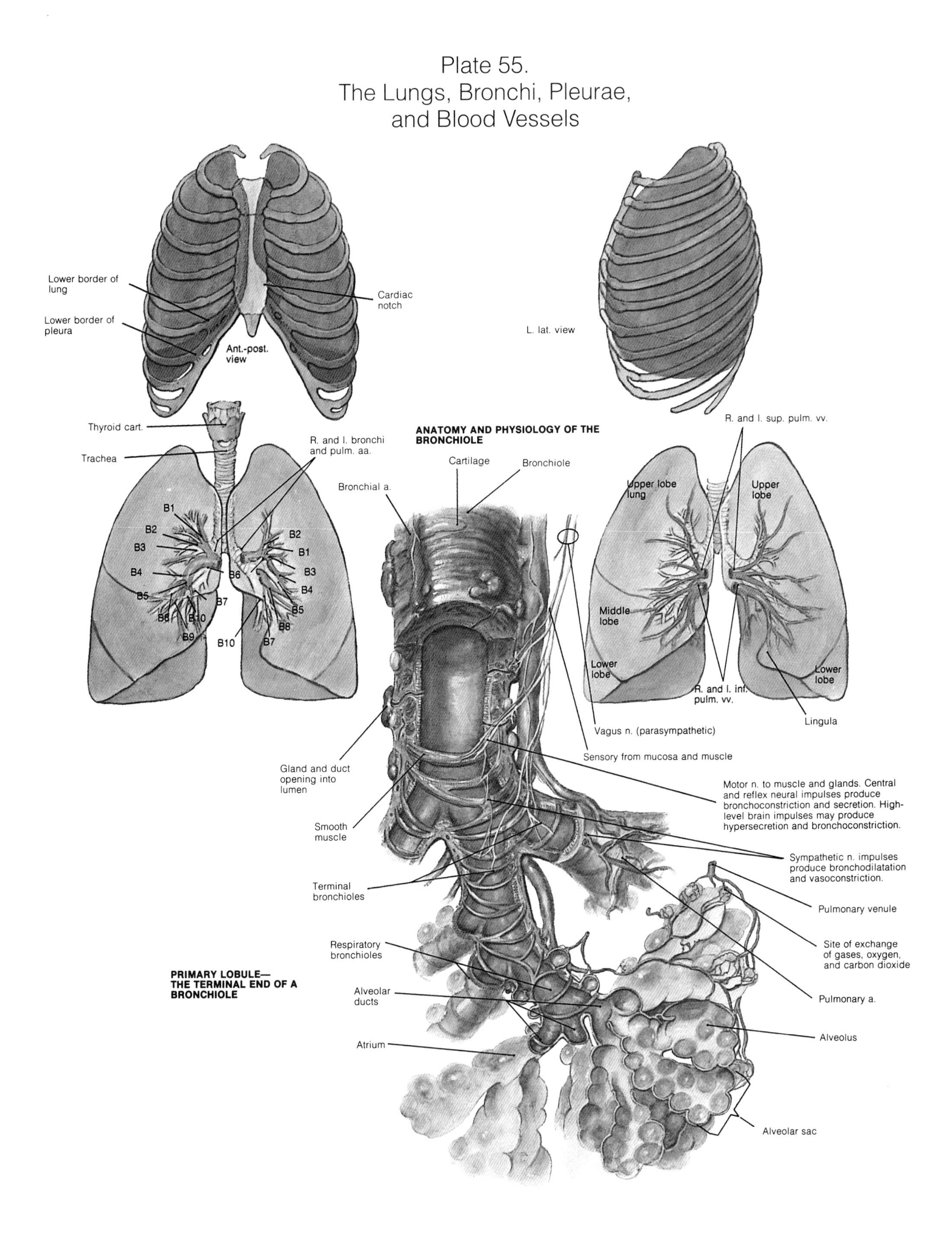

Plate 56.
Lobes of the Lungs

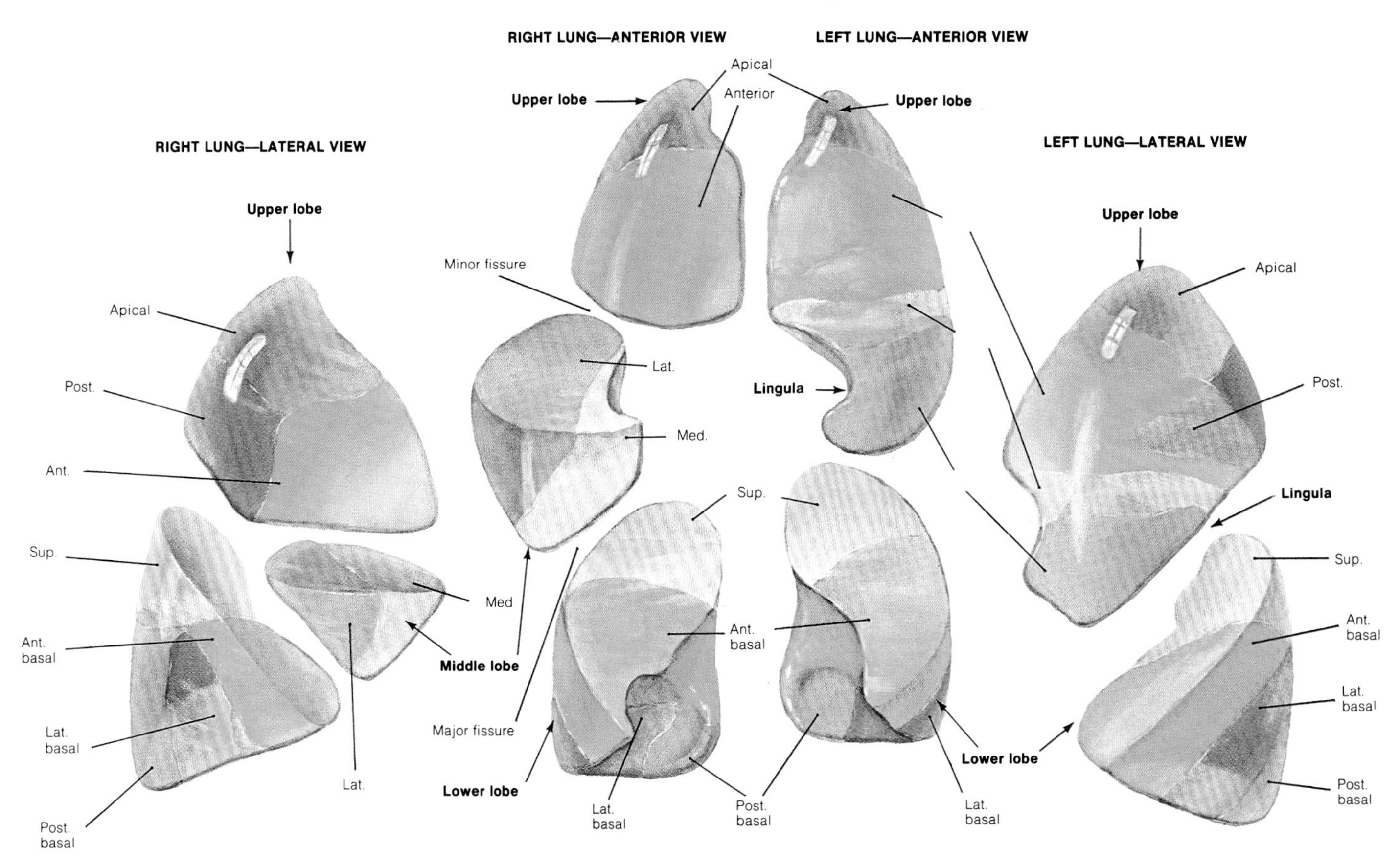

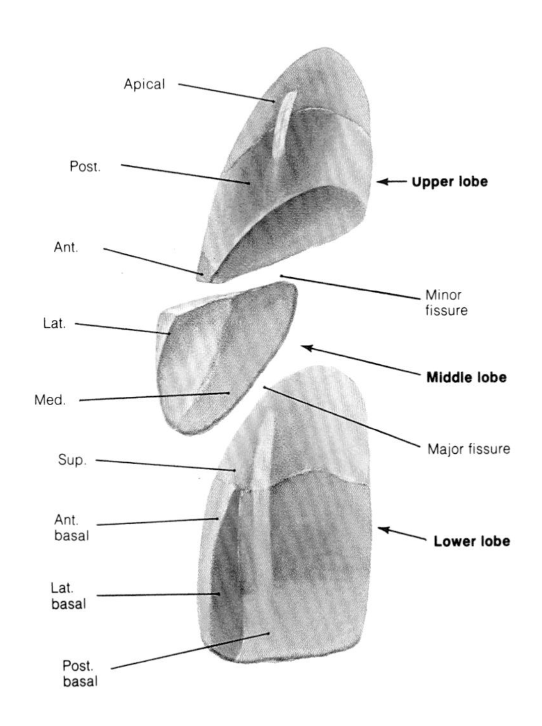

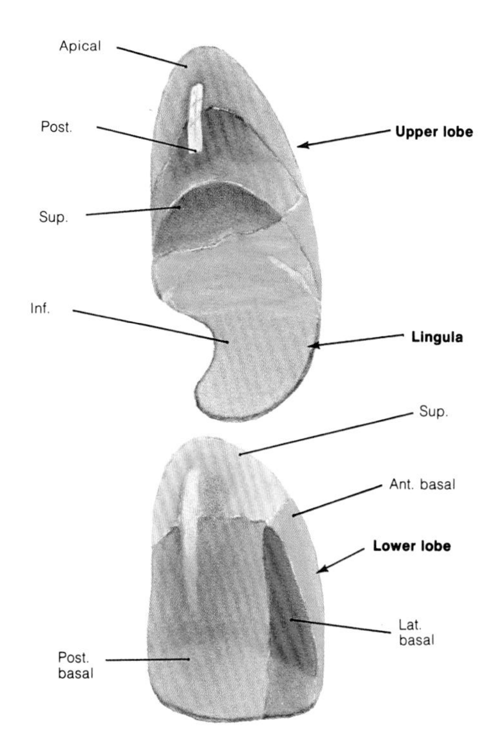

from the alveoli has a great effect on local pulmonary blood flow. The distribution of the pulmonary circulation is therefore in a continual state of change as the result of postural changes, exercise, and other factors. With work or other exercise, the exchange of respiratory gases is increased by more active pulmonary ventilation and an increase in pulmonary arterial blood flow. For this gaseous exchange to be efficient, changes in regional blood flow must match the changes in ventilation. In most diseases, whenever regional ventilation is impaired the pulmonary arterial blood flow to the involved zones decreases as the result of arteriolar constriction. In some diseases, however, the disease is at a level where perfusion and ventilation are no longer matched and cyanosis results. In other diseases, perfusion is reduced without impairment of ventilation. Of these, pulmonary embolism is the most common. In most patients, such emboli are larger than the 2-mm diameter of the segmental pulmonary arteries, and perfusion defects corresponding to specific lung segments can be seen. Diagnostic specificity can be increased if attention is paid to the size, shape, and location of perfusion defects, as seen in lung scans or pulmonary arteriograms.

Although dyspnea with exercise is the most common and perhaps the most sensitive indicator of lung disease, measurement of properties of the lungs such as size, volume of the air spaces, expansibility (compliance or airway resistance), ventilatory ability (vital capacity, maximal breathing capacity) and rate of gas exchange (diffusing capacity), together with regional measurements of ventilation and perfusion, give us a more detailed picture of the gas-exchange function of the lungs. In many cases identification of the pattern of abnormalities can help elucidate the probable cause of the disease in a given patient, help predict what is likely to happen to him, and provide the basis for more rational treatment.

Other functions of the lung are also of great importance: these include coughing, mucociliary activity, and alveolar phagocytosis. These can be examined with a variety of techniques, many of them involving direct measurements with radioactive tracers. These help in two ways: (1) to define abnormalities of individual lung functions in precise quantitative terms; and (2) to locate the areas of function in terms of segments and lobes at the gross anatomical level, or at the microscopic level, at the alveolar, interstitial, bronchial or vascular levels. Such information is useful in planning treatment and in objective evaluation of the results of treatment.

The Gastrointestinal Tract

Thomas R. Hendrix, M.D.
George D. Zuidema, M.D.
John J. White, M.D.

EMBRYOLOGY

A knowledge of the embryologic development of the abdominal viscera is basic to an understanding of the normal relationships that these viscera assume as well as necessary for explaining the congenital malformations that are encountered with reasonable frequency. The embryonic gut is a simple tubular structure that lies suspended from the ventral and dorsal walls of the developing embryo by double-layered ventral and dorsal mesenteries. Fairly early in embryologic development the caudal portion of the ventral mesentery disappears, and there is free communication between the right and left portions of the abdominal cavity. The cranial portion of the ventral mesentery persists, however, and is the site for the development of the liver. The portion of ventral mesentery lying between the liver and the ventral abdominal wall persists as the falciform ligament. The lower free margin of this ligament contains the obliterated left umbilical vein, the ligamentum teres. The persistent portion of ventral mesentery between the liver and stomach becomes the gastrohepatic ligament, or lesser omentum. This extends to the duodenum and at its lower margin contains the common bile duct, hepatic artery, and portal vein.

The portion of the original dorsal mesentery that suspends the stomach is termed the mesogastrium; that part suspending the jejunum and ileum is termed dorsal mesentry; and the portion supporting the colon is the mesocolon. The spleen develops in the cranial portion of the dorsal mesentery and the vascular supply to the foregut arises partially from the celiac axis. The branches of the celiac axis lie between the layers of dorsal mesentery. The hepatic artery supplies the liver, the left gastric artery supplies the stomach, and the splenic artery supplies the spleen. The persistent dorsal mesentery between stomach and spleen is termed the gastrosplenic ligament, and the mesenteric attachment between the spleen and the midline is the splenoaortic ligament.

The foregut undergoes a 90-degree rotation to the right about the longitudinal axis of the stomach. As a result of this rotation the dorsal border of the stomach, the greater omentum, and the spleen become located in the left upper quadrant. A remnant of the splenoaortic ligament becomes the splenorenal ligament.

The foregut undergoes a second phase of rotation in relation to the horizontal axis of the stomach as a result of the unequal rates of growth of the rapidly expanding dorsal mesentery and the slowly growing ventral mesentery. The ventral mesentery fixes the location of the lesser curvature of the stomach while the greater curvature continues to expand. This brings the pyloric end of the stomach upward and to the right where it becomes fixed in position. The dorsal mesogastrium, by its displacement, creates a large sac posterior to the stomach termed the *omental bursa.* The dorsal mesentery is now named greater omentum. A small opening between the greater peritoneal cavity and the omental bursa persists as the foramen of Winslow. The boundaries of this opening include the caudate lobe of the liver above; the duodenum below; the hepatoduodenal ligament containing hepatic artery, common bile duct, and portal vein in front; and the inferior vena cava behind.

The greater omentum is formed by the fusion of the anterior and posterior layers of the original dorsal mesentery. The area that persists posterior to the transverse colon constitutes the transverse mesocolon; the anterior layers constitute the gastrocolic ligament.

Rotation of the Midgut

During embryologic development the primitive tubular midgut is suspended by a dorsal mesentery which divides posteriorly over the retroperitoneal area and con-

Plate 57.
Intestinal Rotation (Fetus)

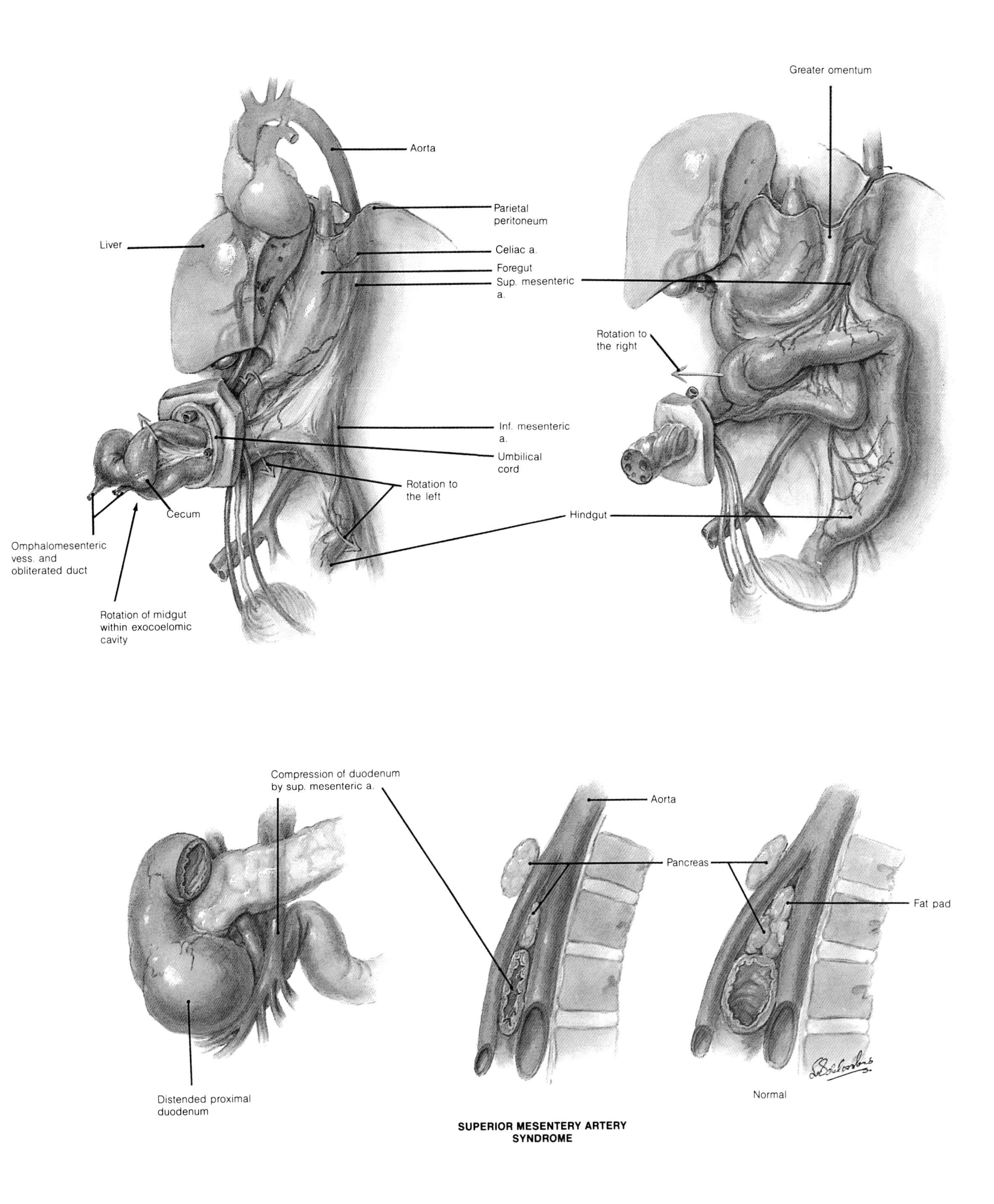

SUPERIOR MESENTERY ARTERY SYNDROME

tinues over the posterior abdominal wall as parietal peritoneum. The midgut extends from the duodenojejunal junction to midtransverse colon. Its blood supply is by the superior mesenteric artery which lies between the two layers of dorsal mesentery and extends into the yolk sac as the vitelline artery. The superior mesenteric artery serves as the axis of rotation of the midgut. Rotation occurs in three stages: the first consists of physiologic herniation into the exocoelomic cavity; in the second phase the viscera return to the developing peritoneal or coelomic cavity; the third stage consists of fixation of the viscera into their final locations. Both rotation and mobilization of the intestinal tract occur during all three stages. During the first stage the midgut undergoes rapid elongation. This growth extends to the point at which the coelomic cavity is no longer capable of accommodating it. Consequently, it herniates through the umbilical orifice into the primitive umbilical cord. The superior mesenteric artery is the major structure within the herniated loop which is attached at its other end to the yolk stalk. During this exocoelomic phase the midgut undergoes a 90-degree counterclockwise rotation which throws the cranial portion of the loop to the right and the caudal portion to the left.

As the coelomic cavity enlarges, the midgut returns, retreating from its exocoelomic location. This phase is extremely important, for a further counterclockwise rotation of 180 degrees occurs. The superior mesenteric artery remains as the axis of rotation during this phase of reduction. The proximal portion of the cranial loop is reduced first, passing under the superior mesenteric artery. The caudal loop is the last to reenter the abdominal cavity after completion of the rotation. Upon completion of this stage the cecum and proximal colon (distal portion of caudal loop) become situated in the right upper quadrant of the abdomen. This stage completes the rotation of the midgut.

During the final, or fixation, stage the cecum descends from its subhepatic location to its normal position in the right side of the peritoneal cavity. The dorsal mesentery of the cecum and ascending colon becomes obliterated and fixed to the parietal peritoneum in the right flank. The transverse colon remains suspended from the transverse mesocolon from the hepatic flexure to the splenic flexure. The mesentery of the small bowel extends from the duodenojejunal junction at the ligament of Treitz to the ileocolic valve; the mesenteric base is attached along an oblique line extending downward from the ligament of Treitz toward the right lower quadrant. On the left side the descending colon loses its peritoneum and becomes fused along the left lateral posterior peritoneal wall, and only the sigmoid colon retains a mesentery.

A variety of congenital malformations of the abdominal viscera may result from abnormal patterns of rotation, particularly involving the midgut. The anatomical relationships of the umbilical region of the fetus at term are also shown. At this stage the ductus arteriosus is patent, as are the sinus venosus and umbilical vein, the urachus and the hypogastric arteries. The umbilical veins and the hypogastric arteries are combined as the omphalomesenteric vessels within the umbilicus. These structures become obliterated as does the sinus venosus as the fetus begins extra-uterine life. Failure of these structures to obliterate leads to a variety of congenital malformations, samples of which are described. An omphalocele (not shown) results when rotation fails to continue beyond the first stage and the midgut remains arrested within the exocoelomic cavity. The loop of bowel is covered by amnion and the abdominal wall may be incompletely developed so that a massive defect may be present. The hernia may consist of a single loop of bowel or may contain virtually the entire intestinal tract, liver, spleen, and pancreas.

Several important anomalies may be produced by irregularities of rotation and fixation of the midgut. Failure of the stage of fixation (stage three) may result in the cecum occupying a subhepatic location. If the visceral peritoneum fails to fuse with the posterior parietal peritoneum, the entire cecum and ascending colon remain free on a mesentery and mobile. This makes cecal volvulus possible.

The most common anomaly of intestinal rotation is encountered when the appendix becomes fixed under the cecum during its descent into the right lower quadrant. The appendix then persists in a retroperitoneal or retrocecal location (not shown).

In the early human embryo the umbilical loop of the primitive gut communicates freely with the yolk sac by way of the omphalomesenteric duct. As development proceeds, this duct normally becomes occluded and later disappears entirely. All or any part of the duct may, however, persist. If the duct remains attenuated, but with an intact lumen, an umbilico-intestinal fistula results. The remnant of the duct may be very short so that the ileum appears to be attached to the undersurface of the abdominal wall, or it may be several inches in length. A variety of anomalous arrangements may result, ranging from an open communication with prolapse of the bowel through the opening, producing intestinal obstruction, to a very narrow lumen which is productive of only a small mucous discharge.

If the midportion of the duct remains present, an intestinal cyst may occur. This cyst may communicate with the umbilicus or the ileum, or only the middle portion may persist as a blind cyst.

If the proximal portion of the omphalomesenteric duct remains persistent, a blind diverticulum attached to the ileum results. This is termed Meckel's diverticulum and is a very common anomaly of the intestinal tract. It is usually located within a meter of the terminal ileum and may or may not have an associated mesentery.

In addition to anomalies in the rotational and fixation

Plate 58.
Small Intestinal Malformations—Embryologic Considerations

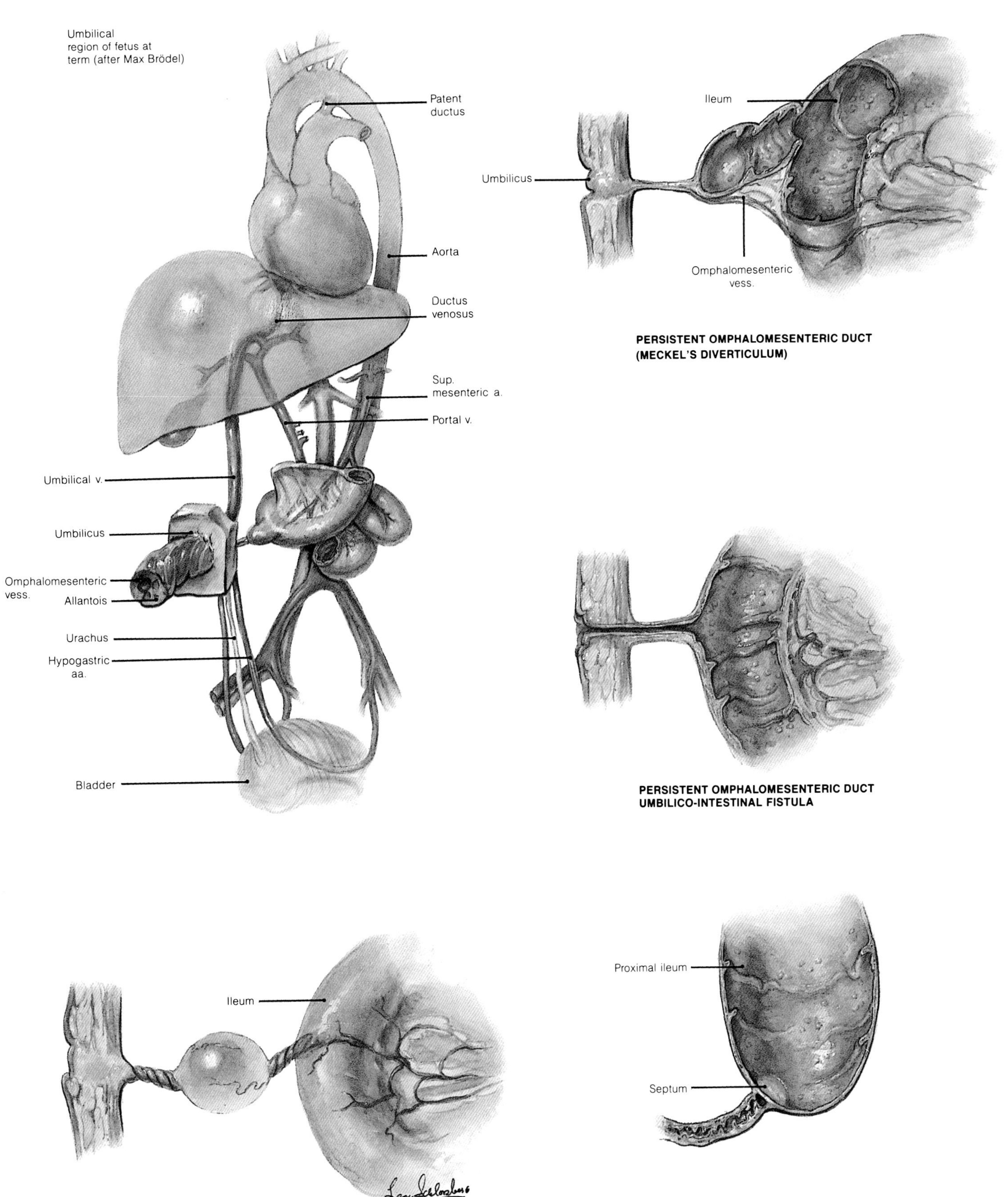

PERSISTENT OMPHALOMESENTERIC DUCT (MECKEL'S DIVERTICULUM)

PERSISTENT OMPHALOMESENTERIC DUCT UMBILICO-INTESTINAL FISTULA

INTESTINAL CYST

INTESTINAL ATRESIA

phases of midgut development, anomalies may also be encountered as a result of failure of lumen development, such as atresia, stenosis, or duplication of the bowel lumen. During the sixth week of embryonic life the epithelial lining of the gut proliferates rapidly. This results in temporary obliteration of the lumen to varying degrees. Later, vacuoles appear within the solid epithelial cord of the alimentary tract. As the vacuoles coalesce, normal lumen patency occurs. Failure of coalescence of the vacuoles, however, may result in persistence of a transverse septum and obliteration of the lumen, resulting in intestinal atresia. Stenosis may be the result of a variant in this same process in which the transverse septum undergoes incomplete resolution.

An alternate theory to explain the development of intestinal atresia or stenosis is based on the hypothesis that local abnormalities in the blood supply occur, producing small areas of vascular insufficiency, leading to abnormalities in development. An intestinal duplication, with development of an enteric or enterogenous cyst, results when the vacuolation produces a longitudinal septum. Duplication cysts may occur in any portion of the alimentary tract, and they usually do not communicate with the intestinal lumen.

The urachus is a midline fibrous cord extending from the apex of the bladder to the umbilicus. During embryonic life, this structure is the allantoic stalk. If the lumen is persistent, the result is a fistula between the umbilicus and the urinary bladder.

Another variation that may be encountered is a state in which the intestinal tract returns to the coelomic cavity but rotation beyond the initial 90-degree turn is arrested. This results in placing the cranial limb of the embryonic midgut in the right side of the peritoneal cavity and the caudal limb in the left side so that in the fetus the small intestine lies entirely on the right side of the abdomen, with the cecum and the entire large bowel located on the left. This anomaly may be completely asymptomatic and detected only when barium studies are performed for some other reason later in life. It may be particularly troublesome if appendicitis develops in a patient with this aberrant condition.

In rare instances reversed rotation (clockwise) may occur during the stage of reduction into the coelomic cavity. In this situation the duodenum lies anterior to the superior mesenteric artery and the transverse colon arises behind. As the root of mesentery becomes fixed, the transverse colon may be obstructed by the superior mesenteric artery.

If intestinal rotation is arrested upon the completion of stage two, fixation of the mesentery fails to occur, and the entire midgut loop hangs from a narrow pedicle based on the superior mesenteric vessels. This arrangement predisposes to volvulus of the bowel around the superior mesenteric vessels. In the worst situation this may result in extensive intestinal infarction. More often the patient may present with symptoms of partial or complete duodenal obstruction dating from birth.

Internal hernias may result if the returning cranial loop of midgut rotates into the mesentery of the caudal loop rather than into the general peritoneal cavity. With fixation of the mesentery loop the bowel may become imprisoned beneath the mesentery, and a paraduodenal hernia may result.

ANATOMY

The mouth and pharynx serve as the entrance to both the respiratory and gastrointestinal tracts. The oropharyngeal musculature is striated and is innervated by the cranial nerves. Its complex, stereotyped activities are coordinated by centers in the brain stem (medulla oblongata), e.g., the swallowing center, respiratory center, vomiting center, etc. The mouth and pharynx are lined with stratified squamous epithelium and lubricated by the secretion of the simple buccal glands and the complex salivary glands, the parotid, submaxillary, and sublingual glands.

The esophagus is the conduit connecting the pharynx to the stomach. Its muscular coat is composed of two layers, an outer one with longitudinally arranged fibers and an inner one with circularly arranged fibers. The upper one-third of the esophageal musculature is striated muscle, whereas the lower two-thirds are smooth. The extrinsic innervation of the esophagus is from the esophageal plexus. The esophagus is also lined with stratified squamous epithelium. It passes down the posterior mediastinum through the esophageal hiatus in the diaphragm, to enter the stomach in the abdomen.

The arterial blood supply of the esophagus is segmental in nature, with the upper portion derived from the inferior thyroid artery, the midportion arising from esophageal branches of the intercostal and bronchial arteries, and the lowermost part supplied by ascending branches of the left gastric artery. The venous drainage is via the azygos vein into the superior vena cava above, and by the coronary vein, entering the portal vein below.

The stomach is fixed at its two poles, the esophagogastric junction and the gastroduodenal junction. The proximal stomach, or fundus, lies high in the left upper quadrant of the abdomen under the diaphragm. The body of the stomach courses anteriorly and to the right, where it becomes the gastric antrum, which crosses the spine anterior to the pancreas to join the duodenum, which is located retroperitoneally. The stomach has three coats of smooth muscle: an outer longitudinal, a middle circular, and an inner oblique coat. The stomach is lined by two types of glandular mucosa, which are arranged in folds, or rugae. The proximal two-thirds is the fundic mucosa, with glands

Plate 59. Esophagus—Arteries and Nerves

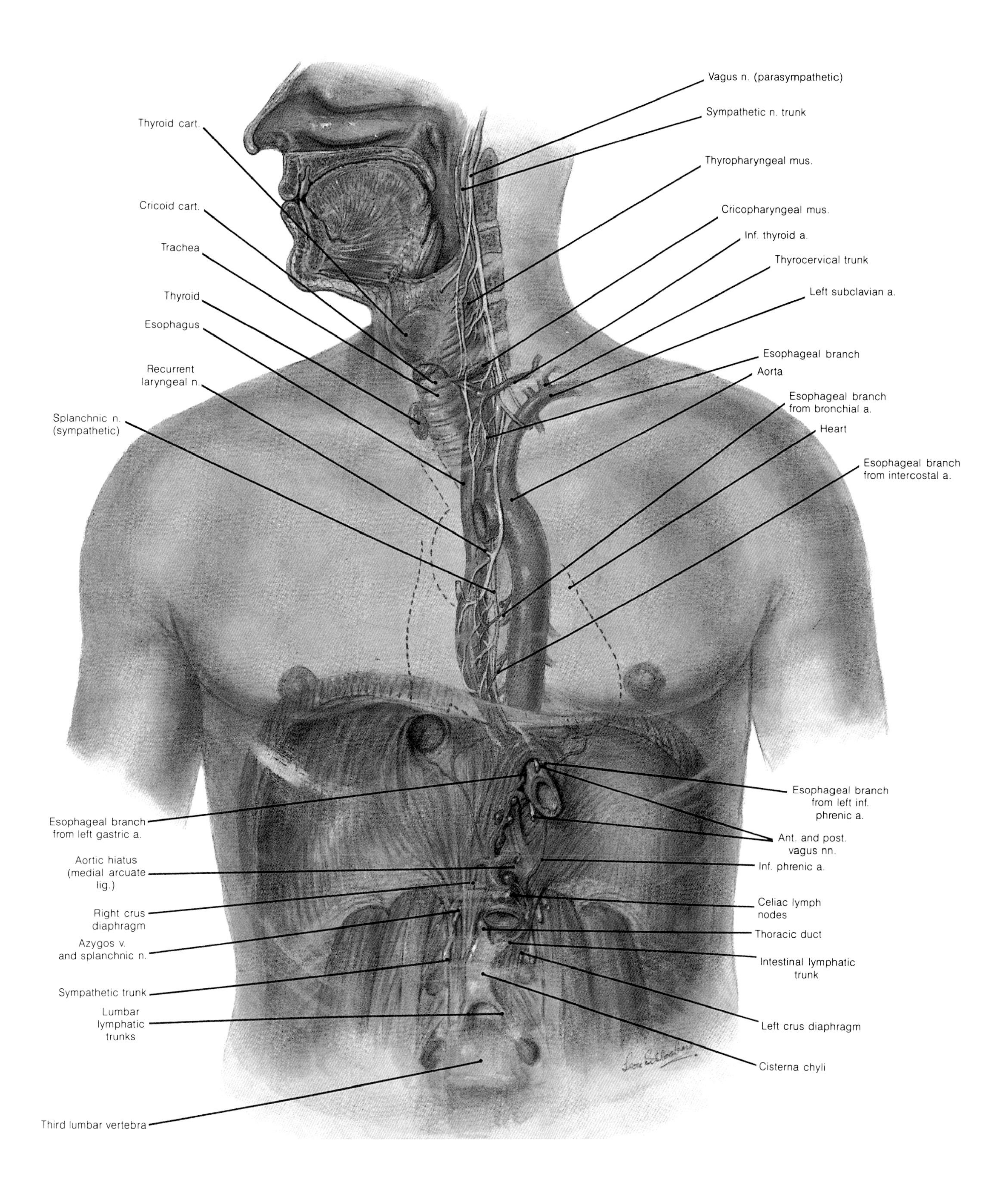

containing the acid-producing parietal cells and the pepsin-producing chief cells. Antral mucosa lining the distal one-third has glands containing mucin-producing cells and G, or gastrin-producing, cells. The surface of both types of mucosa is covered by mucus-secreting cells.

Extrinsic stimulation of gastric secretion and motility arrive via the vagus nerves, which enter the abdomen along with the esophagus. Vagotomy, the interruption of vagal impulses to the parietal cell gland area and the antrum, is often used in combination with a gastric drainage procedure in the treatment of patients with duodenal ulcer. In a truncal vagotomy, the primary vagal trunks are sectioned. In a selective vagotomy, the major branches to the stomach are cut but the hepatic branches are preserved. In parietal cell vagotomy, or highly selective vagotomy, the hepatic and antral branches are preserved, and only those branches to the fundus and body of the stomach are cut.

The arterial blood supply of the stomach comes from the celiac axis. The left gastric artery supplies the lower esophagus and the lesser curvature of the fundus and body of the stomach. The hepatic artery branch of the celiac axis gives rise to the right gastric artery to the antrum and the gastroduodenum and continues along the greater curvature as the right gastroepiploic artery. The splenic artery gives off the left gastroepiploic artery, which furnishes the blood supply to the greater curvature aspect of the body and fundus. The collateral circulation of the stomach is extensive, and it is usually possible to ligate two of the main arteries without devascularizing the stomach. The venous drainage parallels the arterial supply and enters the portal system. These collateral veins are of clinical importance in the presence of portal hypertension.

The primary venous drainage of the stomach and intestines is the portal vein, formed by the confluence of the superior mesenteric and splenic veins. The inferior mesenteric vein enters the splenic vein a short distance from the origin of the portal vein, and the coronary vein draining the stomach also enters close to the confluence of superior mesenteric and splenic veins. When portal vein obstruction occurs, collaterals develop to the systemic venous circulation through a variety of pathways. The coronary vein connects through submucosal vessels in the esophagus and, hence, from esophageal veins to the azygous system. Increased pressure and flow through this channel causes esophageal varices. The umbilical vein may recanalize and drain to abdominal wall vessels. There are extensive collaterals in the pelvis through the hemorrhoidal system. Collaterals may develop from the base of the mesentery into retroperitoneal veins and, hence, into the lumbar veins. The hepatic veins are very short and enter directly into the vena cava just before it enters the right atrium.

The duodenum, the first portion of the small intestine, is retroperitoneal. It curves around the head of the pancreas, crosses the spine along the lower margin of the pancreas, and reenters the peritoneal cavity at the ligament of Treitz, where it becomes the jejunum. The duodenum receives the acid gastric chyme and the alkaline secretions of the pancreas and liver, as well as the secretion of the many mucosal Brunner's glands. The small intestine, duodenum, jejunum, and ileum all have two muscular coats, an outer longitudinal and inner circular coat. The small intestine is lined with columnal epithelium arranged in tall finger- and leaflike projections, or villi, and short tubular glands, the crypts of Lieberkühn. In addition the mucosa is arranged in circular folds called *valvulae conniventes*. The jejunum and ileum are suspended in the abdomen on the mesentery, which contains the arteries, veins, lymphatics, and nerves serving the intestine. The superior mesenteric artery provides the major blood supply to the jejunum and ileum.

In the right lower quadrant of the abdomen, the ileum enters the medial side of the cecum, the proximal portion of the colon. The appendix, a vestigial structure in man, is located at the dependent, blind end of the cecum. The ascending colon passes from the cecum in the right lower quadrant up to the hepatic flexure under the liver in the right upper quadrant. From this point, the transverse colon crosses to the left upper quadrant, where at the splenic flexure it becomes the descending colon.

As the colon enters the left lower quadrant of the abdomen, it becomes the sigmoid colon. It leaves the peritoneal cavity posterior to the bladder to become the rectum, and opens onto the perineum through the anal canal.

The colonic wall consists of mucosa, submucosa, a circular layer of smooth muscle, a longitudinal layer of smooth muscle and a serosal lining. The external surface of the colon has three characteristic structures: (1) the taeniae coli; (2) the haustral markings; and (3) the appendices epiploicae. The three taeniae coli are longitudinal bands which run the entire length of the colon and represent thickened portions of the longitudinal muscle layer. The haustral markings are sacculations formed in the spaces between the taeniae coli and are separated from each other by the circular layer of smooth muscle. The haustral markings are most prominent when the taeniae coli are contracted. The appendices epiploicae are small sacs filled with fat which are located beneath the serosa.

The blood supply of the ascending and transverse colons is furnished by the ileocolic and middle-colic branches of the superior mesenteric artery; and the descending and sigmoid colons are supplied by the left colic and sigmoidal branches of the inferior mesenteric artery.

The ileocolic artery arises from the right side of the distal superior mesenteric artery and supplies the appendix, cecum, and terminal ileum. The cecum and the

Plate 60.
The Stomach—Neural Considerations

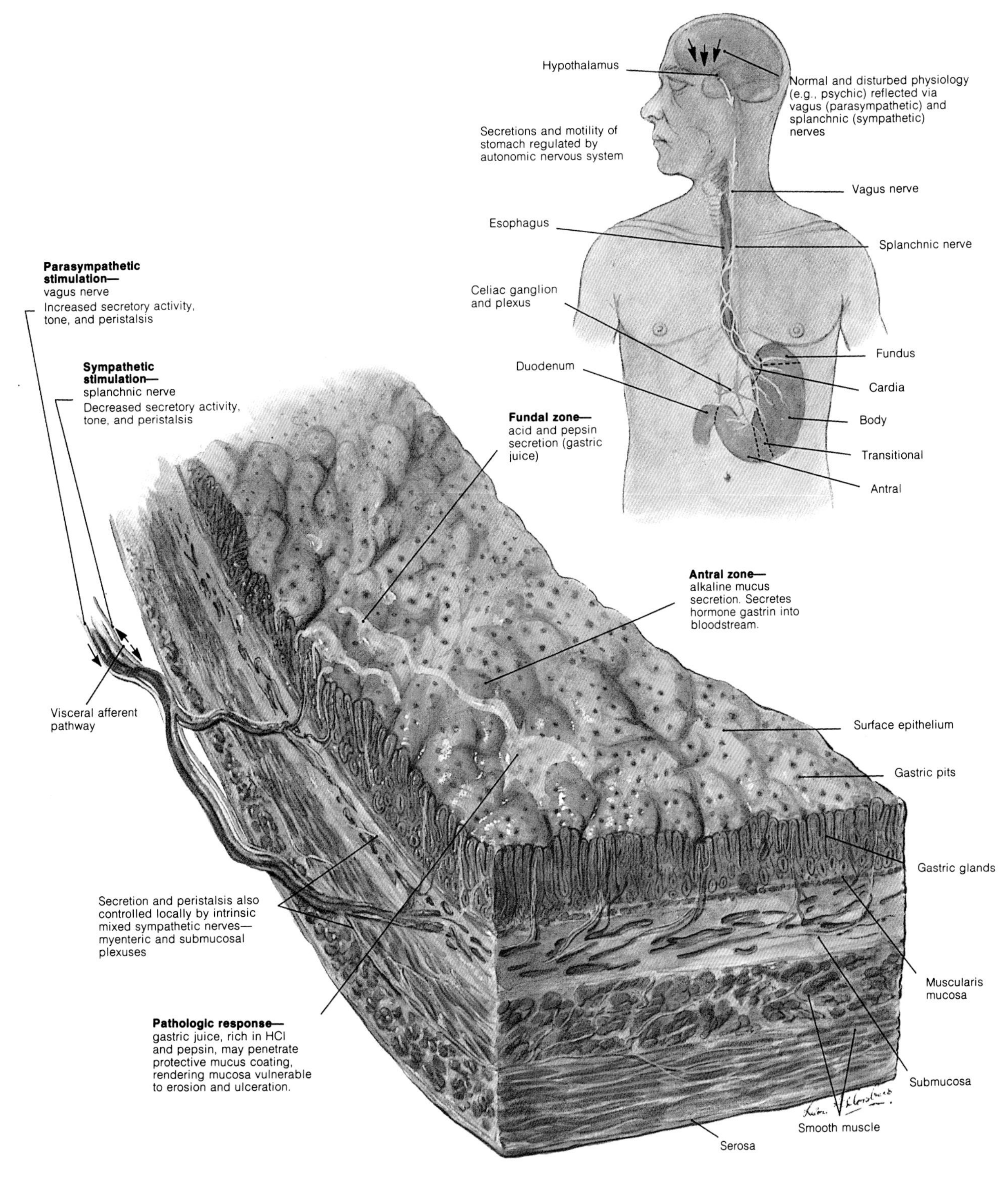

BLOCK OF STOMACH WALL

Plate 61.
Pancreas and Stomach

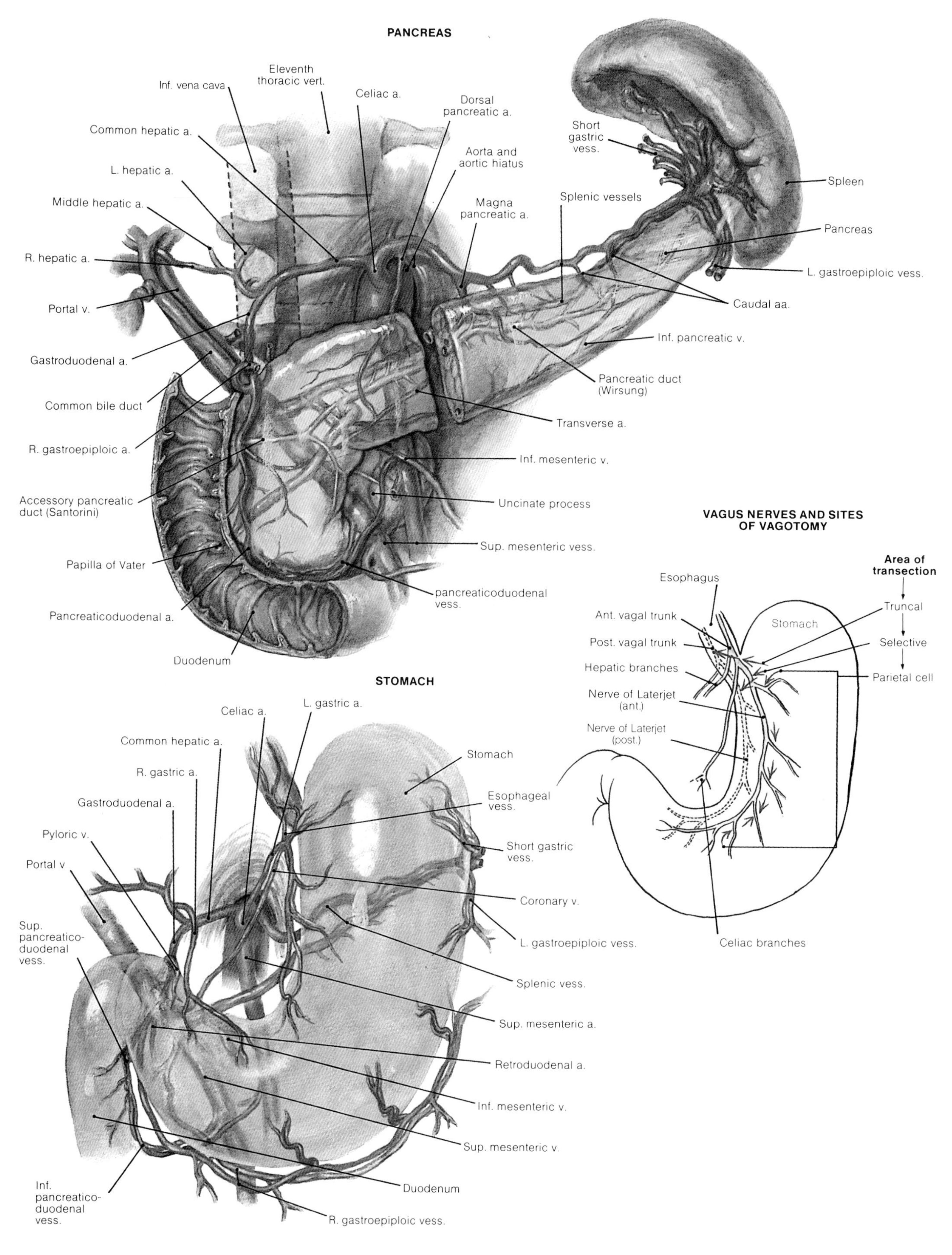

Plate 62.
Esophagus—Liver
Venous Drainage and Collateral Channels

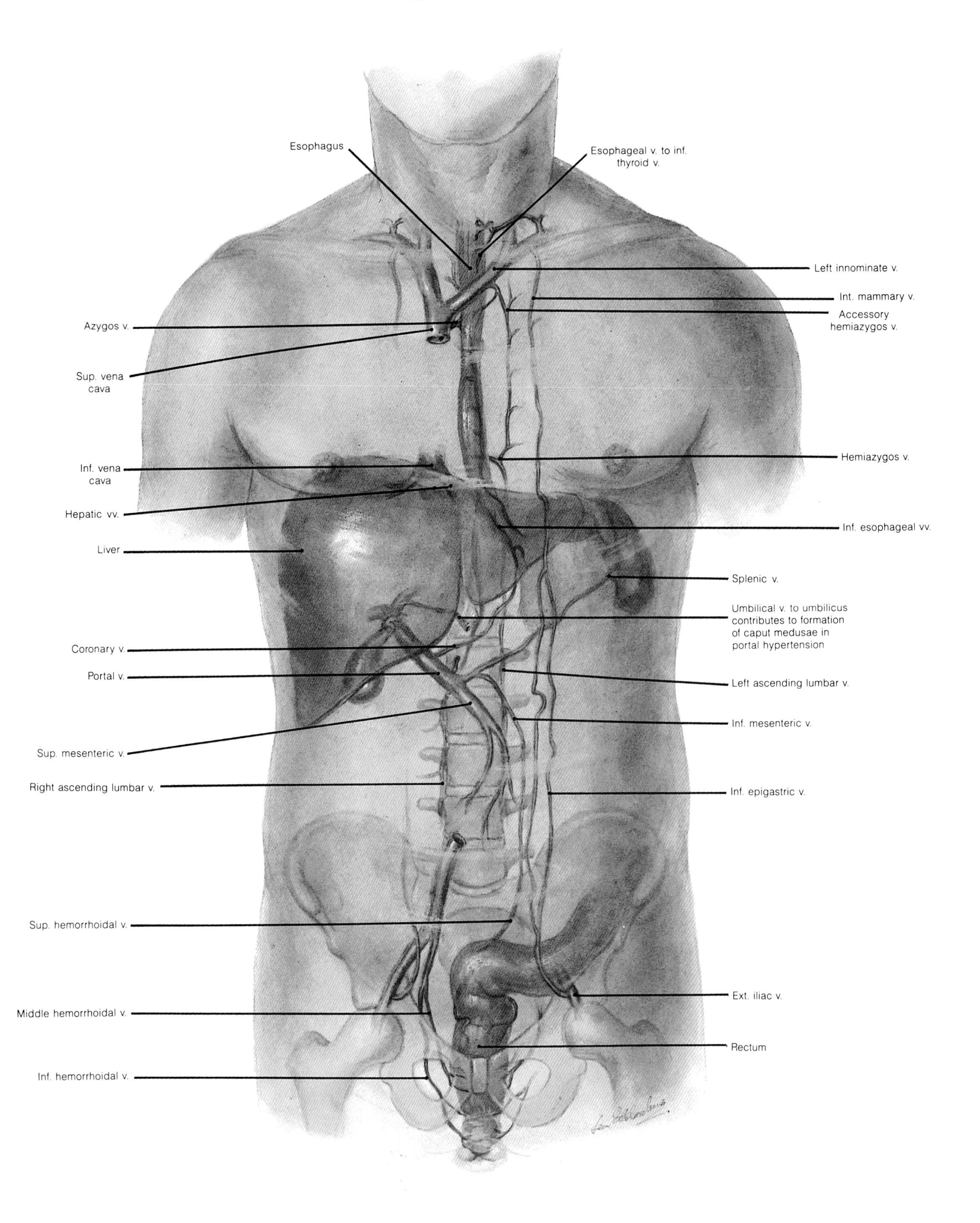

Plate 63.
The Small Intestine

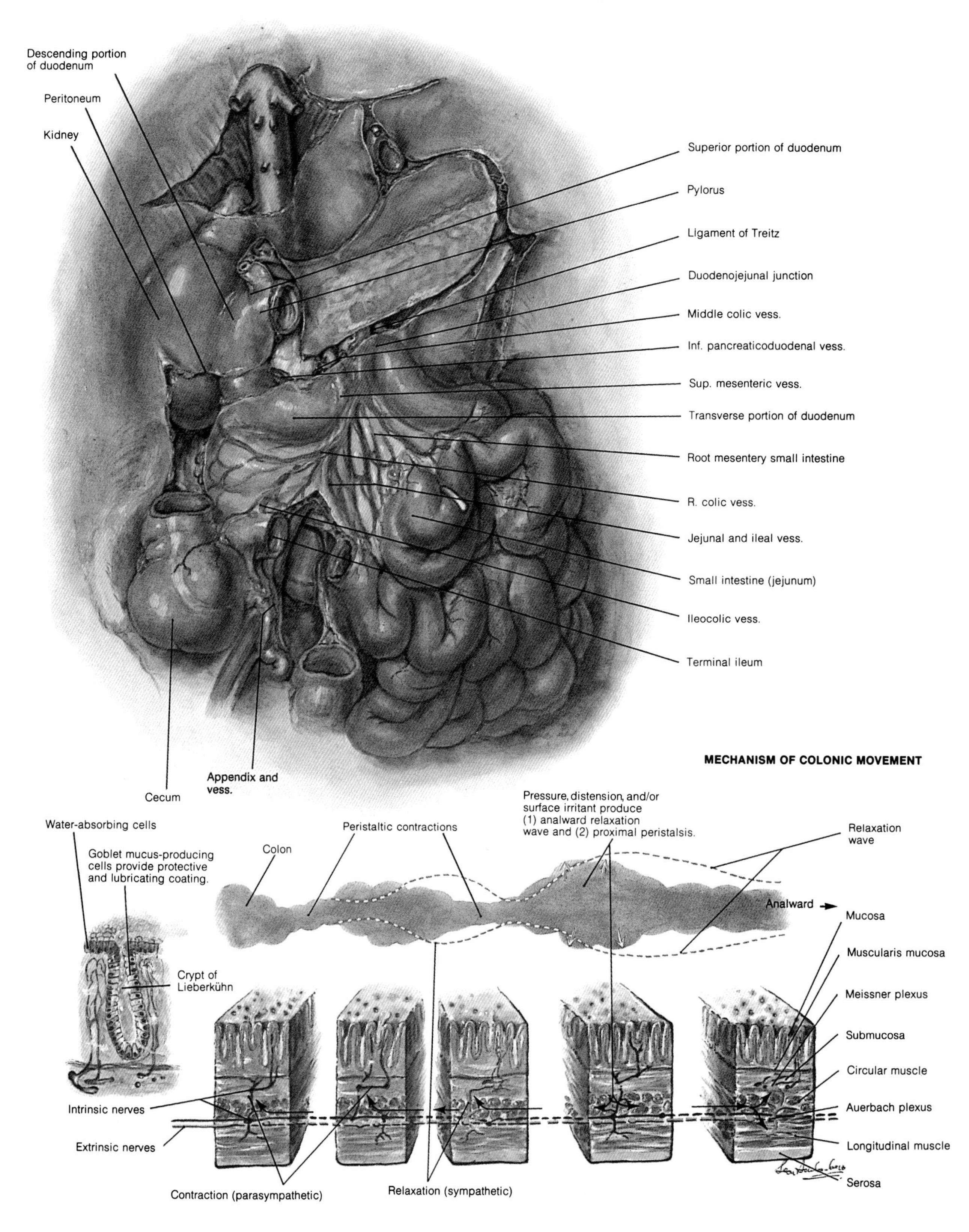

Plate 64.
Colon, Rectum, Anus, and Perineum

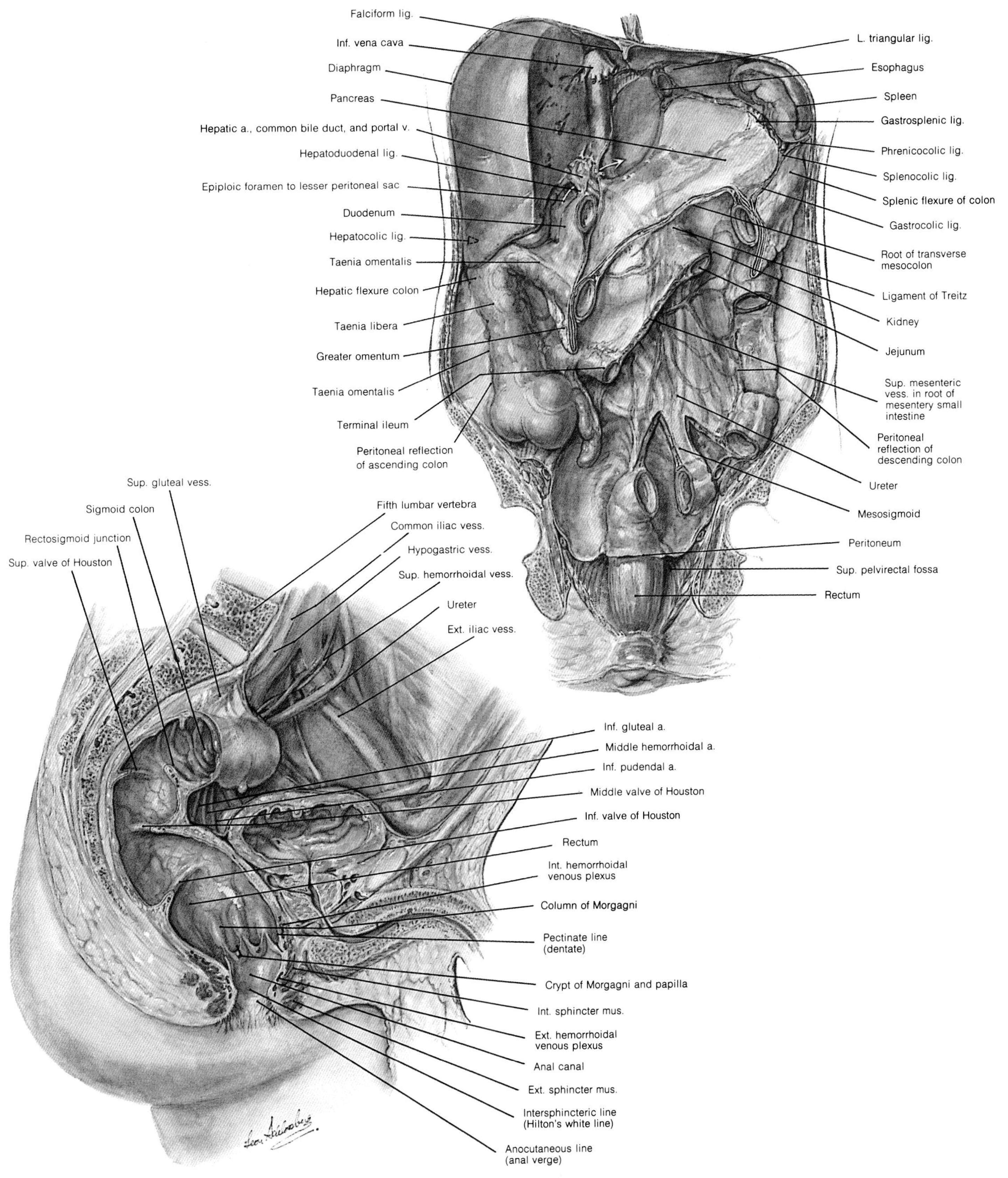

ascending colon receive their blood supply from the right colic artery, which comes off the right side of the distal superior mesenteric artery. The right colic artery has extensive anastomoses with the ileocolic and middle colic arteries via the arcades of the marginal artery. The transverse colon is supplied by the middle colic artery arising from the superior mesenteric artery. This vessel has anastomotic connections with the right colic artery and also with the inferior mesenteric artery via the arcades of the marginal artery. The left colon and sigmoid colon receive their blood supply from the inferior mesenteric artery, which arises from the distal aorta below the renal arteries. The inferior mesenteric artery branches into the left colic artery, which supplies the proximal left colon, and the sigmoid artery, which supplies the distal left colon and sigmoid colon. It terminates as the superior hemorrhoidal artery supplying the proximal rectum. The marginal artery of Drummond consists of anastomotic arcades that connect the ileocolic, the right colic, and the middle colic arteries, which arise from the superior mesenteric artery, to the left colic and sigmoid arteries, which arise from the inferior mesenteric artery.

The proximal rectum is supplied by the superior hemorrhoidal artery. It also receives branches from the middle sacral artery, which arises from the distal aorta. The midportion of the rectum is supplied by the paired middle hemorrhoidal arteries, which arise from the hypogastric, or internal iliac, arteries. The distal rectum and anal canal are supplied by the paired inferior hemorrhoidal arteries, which arise from the internal iliac arteries.

The veins of the colon accompany the arterial branches and have similar names. The ileocolic, right colic, and middle colic veins, the venous drainage of the right colon and transverse colon, drain into the superior mesenteric vein, which joins the splenic vein behind the neck of the pancreas to form the portal vein. The left colic, sigmoidal, and superior hemorrhoidal veins join to form the inferior mesenteric vein, which drains into the splenic vein. The superior hemorrhoidal plexus is located in the submucosa of the lower rectum. The vessels of this plexus form the superior hemorrhoidal veins, which drain into the portal system via the inferior mesenteric vein. Dilatations of these veins are called internal hemorrhoids.

The external hemorrhoidal plexus is located below the pectinate line in the submucosa of the anal canal and the subcutaneous tissues of the perianal area. These drain into the systemic venous system via the inferior hemorrhoidal veins and internal iliac venous systems. Dilatations of these veins are called external hemorrhoids.

Lymphatic Drainage

Collecting lymphatic channels leave the wall of the small bowel at its mesenteric border. There are three groups of regional lymph nodes: a peripheral group located near the mesenteric border adjacent to the bowel wall; a middle group located with the small bowel mesentery; and a central group of nodes which are preaortic in location adjacent to the superior mesenteric artery. The mesenteric lymph nodes comprising these three groups are very numerous and widely distributed.

The lymphatic channels drain into the cisterna chyli which is located deep to the duodenum adjacent to the vertebral body of L2. This lymphatic channel enters the thoracic cavity by passing through the aortic hiatus to become the thoracic duct.

The nerve supply of the small intestine is derived from the superior mesenteric and celiac plexuses of the sympathetic system and from the vagus nerves. The plexuses lie adjacent to the superior mesenteric artery and celiac axis, and the distribution of nerve fibers follows the general course of these vessels. The plexuses contain both sympathetic and parasympathetic fibers. Pain associated with a lesion of the small bowel is referred to areas supplied by the ninth, tenth, and eleventh thoracic nerves. In clinical presentation the pain is usually experienced in the area of the umbilicus, and only occasionally will it spread to involve the lumbar region and the back.

The pancreas is a retroperitoneal structure lying across the upper abdomen. It has both an endocrine function provided by the islets of Langerhans (insulin and glucagon being the most important products) and an exocrine function. The digestive enzymes are produced in the pancreatic acini, and the pancreatic secretion is carried by the pancreatic duct to enter, with the common bile duct, the second portion of the duodenum through the ampulla of Vater.

The exocrine secretions of the pancreas enter the duodenum via the main pancreatic duct at the ampulla of Vater, or the accessory pancreatic duct (Santorini). The gastroduodenal artery lies on the head of the pancreas adjacent to the second portion of the duodenum. This vessel is essential for nutrition of the duodenum, so any injury that disrupts this circulation may produce duodenal necrosis.

The splenic artery and vein run along the posterior, superior surface of the pancreas. The artery gives rise to many small branches to the gland, and the vein receives many small tributaries. These multiple, short vessels make dissection difficult. The tail of the pancreas lies within the splenic pedicle and is easily injured during splenectomy.

The uncinate process is the posterior extension of the head of the pancreas. The superior mesenteric artery and vein course side by side along the uncinate process, deep to the pancreas. Small arterial branches and venous tributaries connect these vessels to the posterior aspect of the pancreas.

The liver, the largest organ in the abdomen, is located

Plate 65.
Colon, Rectum, and Anus—Blood Vessels

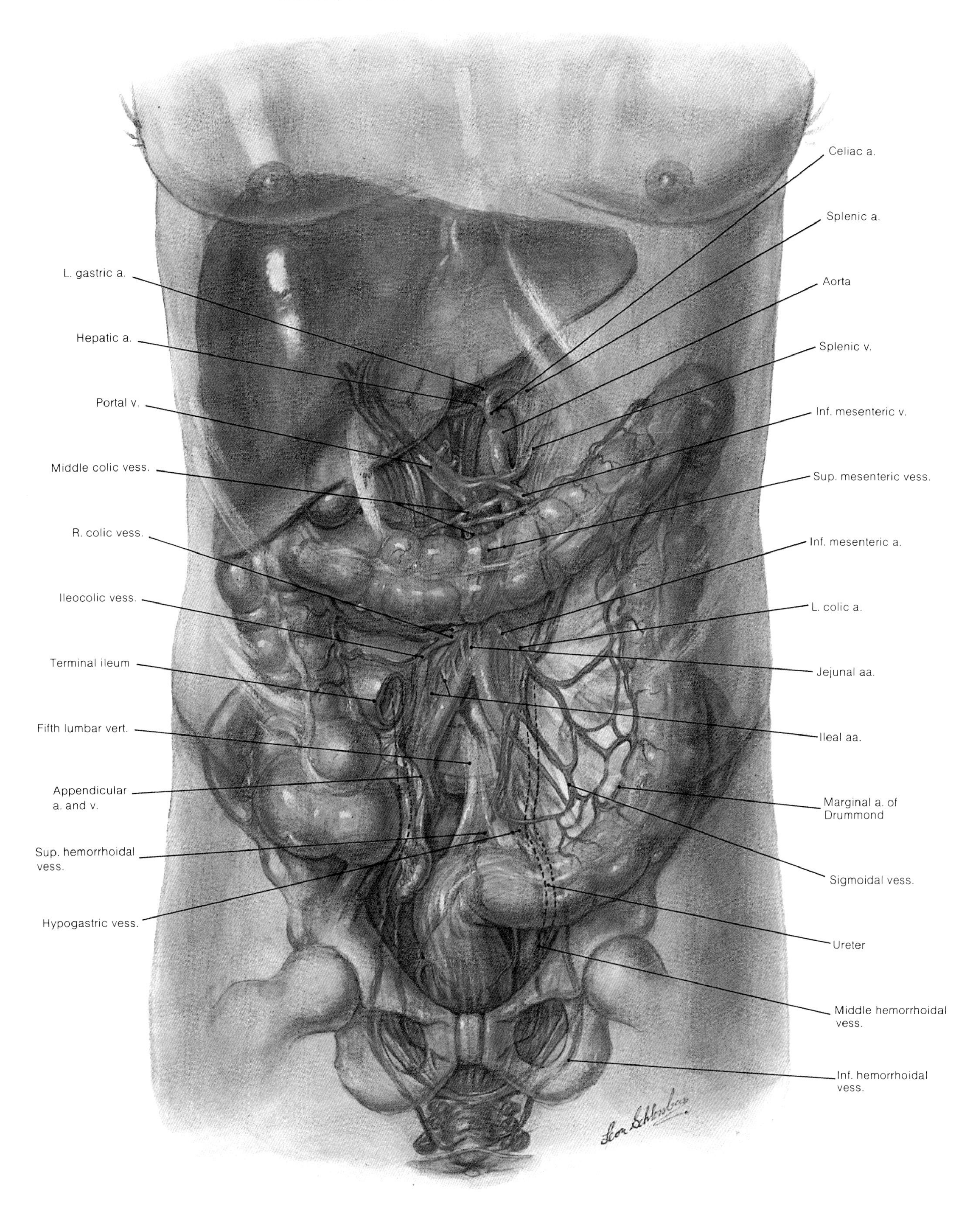

Plate 66.
Colon, Rectum, and Anus—Innervation

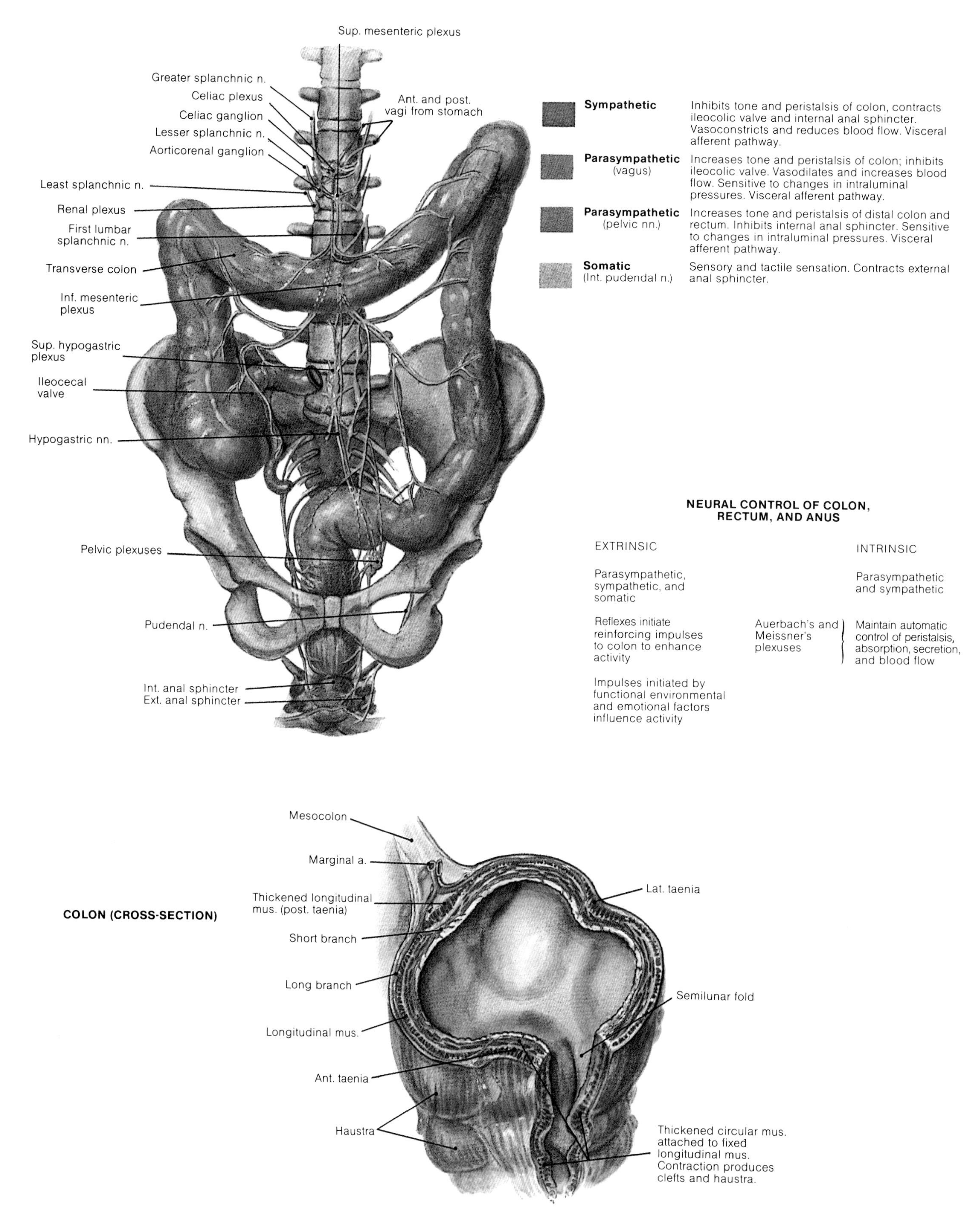

in the right upper quadrant under the diaphragm. The liver has many essential roles in the metabolism of absorbed nutrients: synthesis of proteins; excretion of lipid-soluble materials, including drugs; and the production of bile, which is necessary for the absorption of lipids from the intestine.

The liver has a dual blood supply: arterial blood from the hepatic artery branch of the celiac axis and a much larger volume via the portal vein, which collects the venous blood from the stomach, spleen, intestine, and colon. The portal vein is formed by the joining of the splenic, inferior mesenteric, and superior mesenteric veins posterior and superior to the pancreas. The portal vein branches within the liver to distribute its flow to both lobes and to their segments. The blood leaves the liver to enter the inferior vena cava via the hepatic veins.

The bile formed in the liver is carried by the intrahepatic bile ducts, which coalesce to form the common hepatic duct. The common hepatic duct is joined by the cystic duct to become the common bile duct. The common bile duct passes posteriorly, behind the duodenum, to traverse the head of the pancreas and enter the duodenum through the ampulla of Vater. The main pancreatic duct (Wirsung) also enters the duodenum at this point, either adjacent to the common bile duct or by a common channel.

The gallbladder serves as a reservoir for the bile that is secreted by the liver. It lies on the inferior surface of the liver to the right of the midline and joins the common bile duct via the cystic duct. The cystic artery, a branch of the hepatic artery, furnishes its major arterial supply. Venous drainage is by multiple, small tributaries into the liver bed, joining the portal system.

FUNCTION OF THE GASTROINTESTINAL TRACT

The primary and essential function of the gastrointestinal tract is the absorption of fluids and nutrients. Gastrointestinal secretion and motility serve to prepare the ingested materials for the absorptive process.

The alimentary tract is divided into functional segments by sphincters, which maintain a resting tone greater than the adjacent segments. The oral pharynx and respiratory pathway are separated from the esophagus by the upper esophageal sphincter, and the esophagus is protected from the irritant action of acid gastric juice by the lower esophageal sphincter. These two sphincters, in the resting state, separate the body of the esophagus from the pharynx and airways above and from the stomach below. Both relax in response to swallowing and close after the bolus has passed. The stomach is separated from the duodenum by the pyloric sphincter, which relaxes as the peristaltic wave sweeps down the stomach but allows only a small amount of gastric contents to pass into the duodenum at a time. The pylorus also prevents the reflux of bile into the stomach. Bile salts are damaging to the lining of the stomach and make it susceptible to irritant action of the acid gastric juice. The sphincter of Oddi separates the pancreatic and common bile ducts from the duodenum and prevents reflux of duodenal contents into these glands, but does relax in response to the hormone cholecystokininpancreozymin (CCK-PZ), which causes the gallbladder to contract, emptying the bile into the duodenum. The small intestine is separated from the colon by the ileocecal sphincter. This sphincter prevents the reflux of colonic contents into the small bowel, an important function because contamination of the small intestine with colonic bacteria interferes with its absorptive function. And, finally, the anal sphincters provide control over defecation.

The preparation of food for absorption begins in the mouth, where it is physically broken up by chewing. Digestion of starch is initiated by salivary amylase, which continues until the ingested food is acidified in the stomach. In the stomach, the emulsification process is continued by the churning action of gastric peristaltic waves. In addition, the stomach secretes hydrochloric acid and the protein-splitting enzyme pepsin, which starts the digestion of protein in preparation for its absorption. The peptic digestion of protein is not, however, an essential step in protein absorption. Gastric secretion and motility are controlled by a variety of neural and hormonal influences. The main stimulatory influences are cholinergic fibers in the vagus and the hormone gastrin, which is released from G (gastrin) cells in the antrum. Gastrin release is increased by a variety of stimuli, such as distention of the stomach, protein digestion products, and an alkaline pH in the stomach.

The motility of the stomach, pylorus, and duodenum are nicely integrated so only a small amount of liquefied gastric chyme is delivered to the duodenum at a time. The arrival of this material in the duodenum releases two hormones, secretin and CCK-PZ. Secretin causes the outpouring of a large volume of bicarbonate-rich fluid from the pancreas and, to a lesser extent, increases the flow and bicarbonate concentration of bile. CCK-PZ causes the pancreas to pour out its digestive enzymes; in addition, it causes the gallbladder to contract and the sphincter of Oddi to relax so that bile acids necessary for the solubilization of the products of fat digestion are emptied into the duodenum as the meal is arriving from the stomach. Secretin and CCK-PZ delay gastric emptying until the duodenal contents have been alkalinized to the pH for optimal action of the pancreatic enzymes. Carbohydrates are broken down to disaccharides, which are then hydrolyzed to monosaccharides by enzymes on the brush border of the intestinal cells. These monosaccharides are transported into the cell by a specific carrier mechanism, and leave the intestinal mucosa and enter the general circulation via the mesenteric and portal veins. Proteins are split by

Plate 67.
Liver—Vascular and Biliary Systems

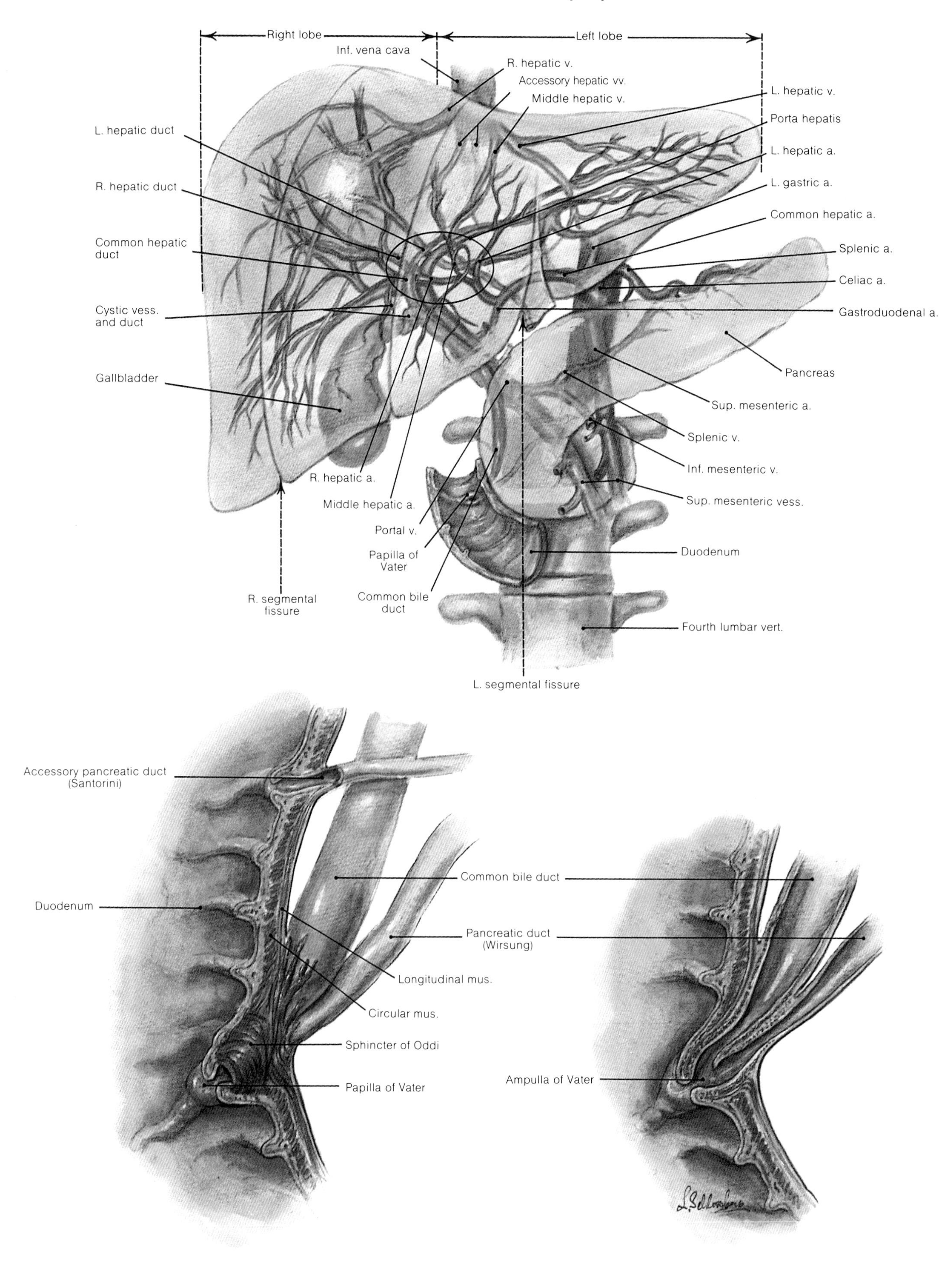

proteolytic enzymes, which are secreted by the pancreas in an inactive form and are only activated through the conversion of trypsinogen to trypsin by the intestinal enzyme, enterokinase. Trypsin in turn activates the rest of the proteolytic enzymes; hence, normally the proteolytic enzymes are inactive while within the pancreas but become active as they enter the duodenum.

The products of protein digestion are peptides, which are taken into the absorbing intestinal cells, where they are split to amino acids, which then leave via the mesenteric and portal veins. Fats are digested by pancreatic lipase to fatty acids and monoglycerides. These digestion products, along with fat-soluble vitamins, are solubilized by the detergent action of bile acids and are carried to the intestinal cell, where they are absorbed. The fatty acids and monoglycerides are resynthesized, again solubilized by the incorporation into a protein envelope, the chylomicron, which leaves the cell and is carried to the general circulation via the lymphatics. Normally, this process of absorption of carbohydrates, proteins, and fats is completed before the food has gone more than one-third of the way through the small intestine. The remainder of the intestine reabsorbs most of the fluid that has entered, either ingested or secreted. The distal small intestine, the ileum, has two specific absorptive functions, first, the absorption of vitamin B_{12}. The absorption of this vitamin is dependent on being bound to intrinsic factor, which is secreted in the stomach and specific transport receptors in the ileum. Secondly, bile acids are specifically absorbed in the ileum and carried back to the liver by way of the portal blood, excreted in the bile, and stored in the gallbladder until the next meal. Normally less than 5 percent of the bile acids are lost into the colon each day.

From one to two liters of fluid are delivered to the colon per day. In the ascending and transverse colon, fluid and electrolytes are reabsorbed and the materials passed. In the descending colon, the unabsorbed food substances, primarily cellulose, desquamated cells, and colonic bacteria are made into solid mass, which is stored and evacuated through the anus by the defecation reflex one or two times a day.

Plate 68.
Colon, Rectum, and Anus—Physiology

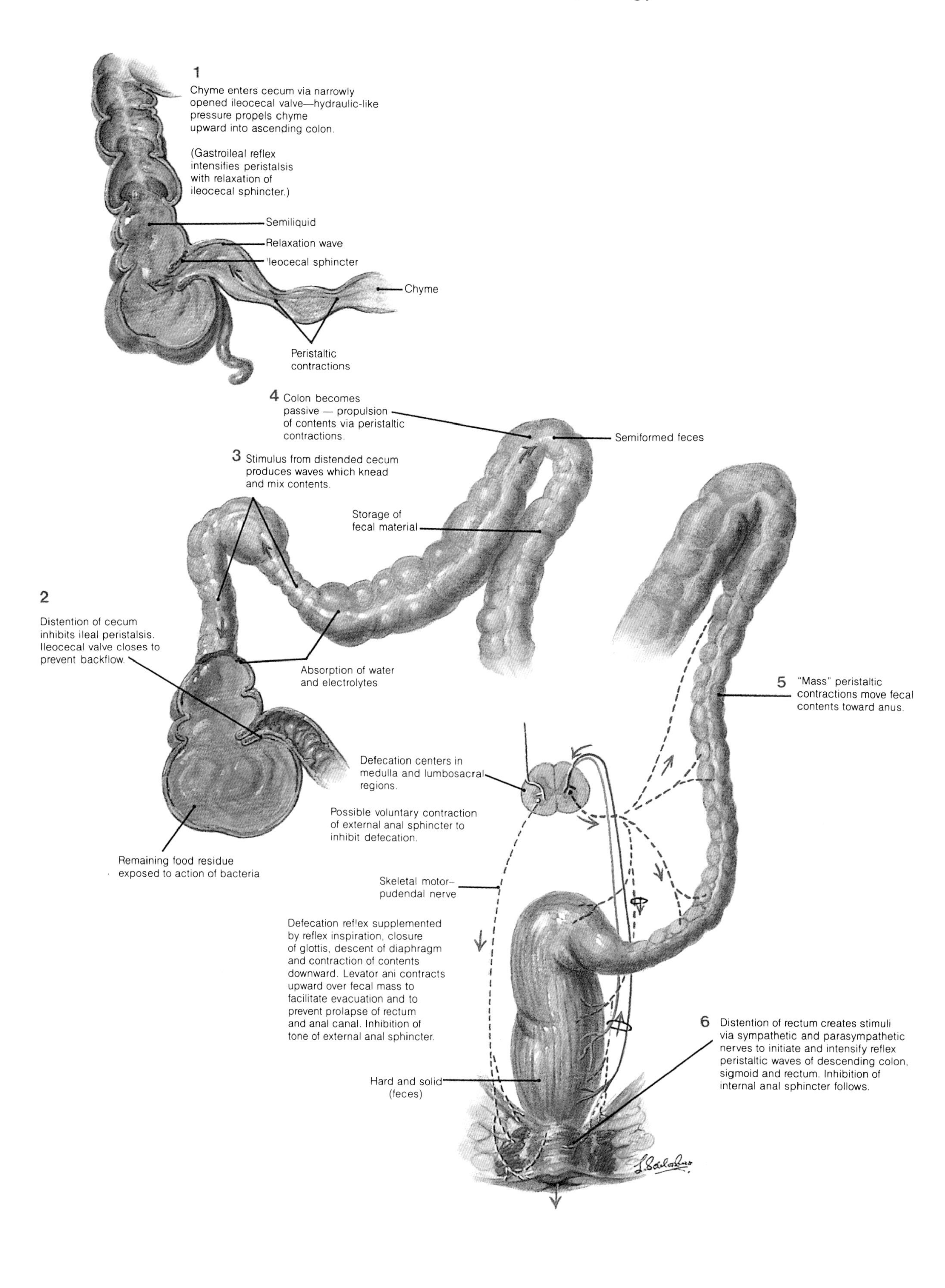

The Liver

Michael A. Choti, M.D.

Significant advances have been made in understanding both the descriptive and the functional anatomy of the liver. Among the various important functions performed by the liver are the secretion of bile, the addition of substances to portal blood, the removal of substances from portal blood, the storage of compounds, and the performance of a large number of important metabolic functions. The liver also has certain unique anatomical features. It has a dual blood supply, deriving oxygenated blood from the hepatic artery and receiving venous blood from the pancreas, intestines, and spleen via the portal vein. Many nutrients and drugs absorbed from the intestine and substances added by the pancreas are presented to the liver in relatively high concentrations. Arterial and portal vein tributaries open into hepatic sinusoids and can communicate with central veins, resulting in direct contact with hepatocytes.

Knowledge of the macroscopic anatomy of the liver is a prerequisite for modern hepatic surgery. Such understanding has helped improve the outcome of major hepatic surgery and has contributed to the advancement of modern surgical techniques, including liver transplantation, intraoperative hepatic ultrasound, and hepatic cryosurgery.

MORPHOLOGIC AND FUNCTIONAL ANATOMY

The classic and now outdated description of liver anatomy regarded the falciform ligament as dividing the right and left lobes of the liver. The portion of the right liver between the gallbladder fossa and the falciform ligament was called the quadrate lobe. The more current and now well accepted classification of liver anatomy more closely parallels functional divisions within the liver. Described first by Cantlie in 1898, it recognized that specific regions of the liver are based on the left and right portal structures. Couinaud (1957) as well as Goldsmith and Woodburne (1957) further refined the functional anatomy, developing a modern system of hepatic segmental nomenclature (Plate 69). The liver is divided into two lobes, the right and the left, separated by the main portal fissure. This line of demarcation goes from the middle of the gallbladder bed anteriorly to the left side of the vena cava posteriorly. Within this plane runs the middle hepatic vein. The portal triad divides into functional units or segments within each lobe of the liver. The right lobe is divided into four segments: anterior-inferior (segment V); posterior-inferior (segment VI); superior-inferior (segment VII); and superior-anterior (segment VIII), with its branching portal structures. The left lobe is divided into the left medial segment (segment IV) and the two segments lateral or to the left of the falciform ligament, the superior component (segment II) and the inferior component (segment III). These two segments are sometimes grouped together as the "left lateral segment." Spatially, the falciform plane is oriented vertically, the plane between posterior and anterior right lobe segments lies horizontally, and the interlobar plane lies diagonally. Segment I, the caudate lobe, is considered unique from a functional point of view; it has independent portal and hepatic venous drainage and is itself divided into left and right processes.

SURFACE ANATOMY

Most evident when viewing the liver from the anterior surface are the *falciform ligament* and the *ligamentum teres,* or round ligament, dividing the left medial from the left lateral segments (see Plate 69). The left portal structures travel from the hilum along the base of the left

Plate 69.
Lobes of the Liver and Their Vascular Supply

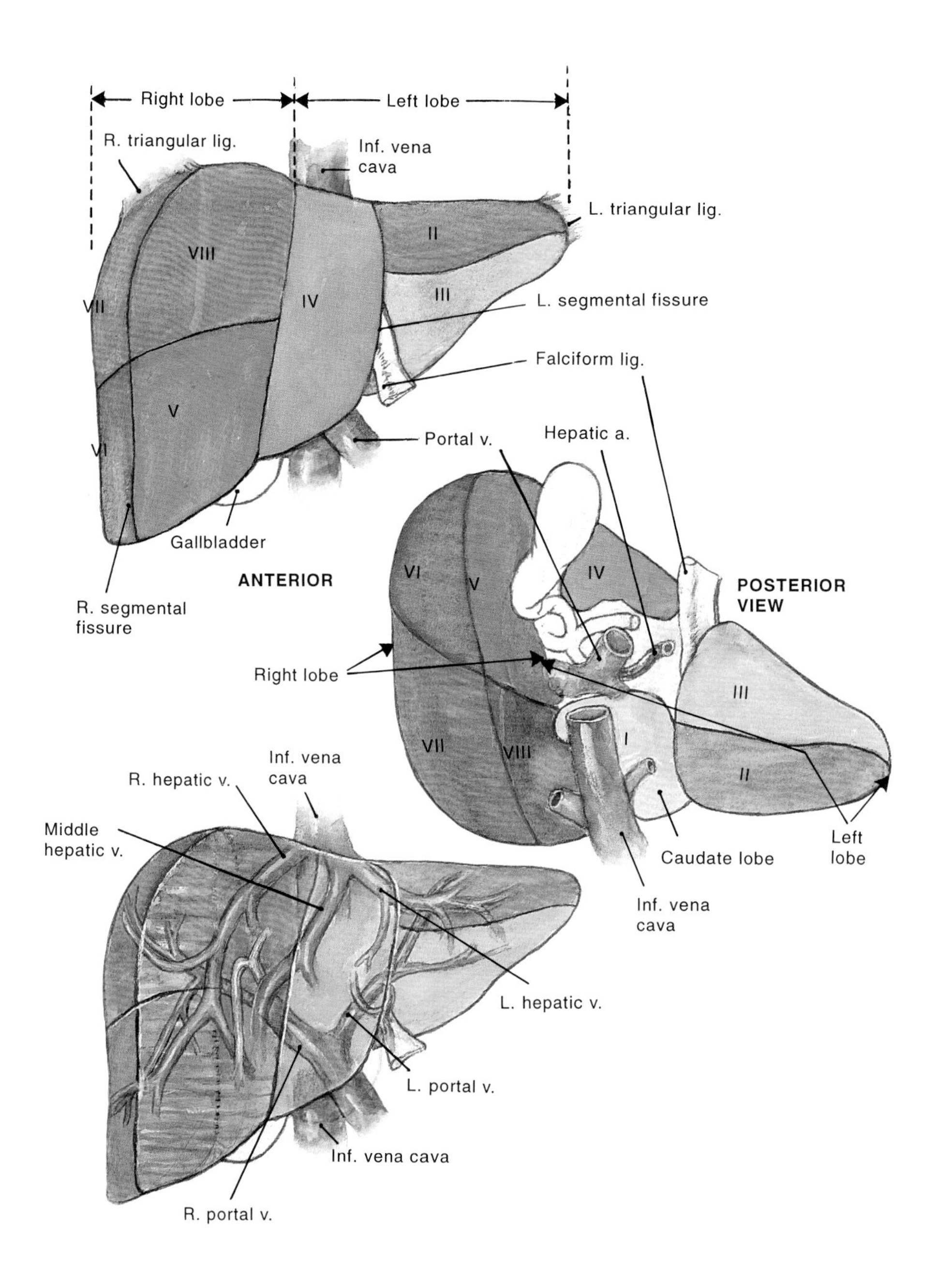

medial segment before entering the left lobe below the ligamentum teres. The falciform ligament tethers the liver to the anterior abdominal wall and divides to form the *left and right triangular ligaments*. The right triangular ligament forms the attachment of the liver to the right diaphragm and the retroperitoneum and incorporates the bare area of the liver. The left triangular ligament fixes the left lateral liver segments to the left diaphragm. The division of the left lateral segments posteriorly from the caudate lobe forms a fissure called the *umbilical fissure,* where the *hepatogastric ligament* attaches to the liver.

The *hepatoduodenal ligament* makes up the hilum of the liver and contains structures including the common hepatic duct, the hepatic artery, and the portal vein. Within this portal triad, the portal vein is typically the most posterior structure and bifurcates closest to the liver. The bile duct lies anterior to the portal vein; and the hepatic artery, while often quite variable in its position, typically is found anterior and to the left within the portal triad.

BLOOD SUPPLY TO THE LIVER

The liver normally receives approximately one-quarter of the total cardiac output and obtains its supply from both the hepatic artery and the portal vein. The hepatic artery normally supplies approximately 25 percent of the total blood flow to the liver. Because hepatic arterial blood is richer in oxygen content, up to 50 percent of the liver's normal oxygen requirement is supplied via the hepatic artery. It typically enters via the hepatoduodenal ligament from the celiac axis as the proper hepatic artery supplying both the right and the left lobes. Variations in extrahepatic anatomy—including a replaced or accessory right hepatic artery from the superior mesenteric artery or a left hepatic artery from the left gastric artery—are common.

The portal vein collects venous outflow from the entire intestinal tract as well as the pancreas and spleen. Seventy-five percent of the total blood flow to the liver, or 90 cc per minute per 100 g of liver weight, is from the portal vein.

HEPATIC VEINS

All blood exiting the liver does so through the hepatic venous system, which carries approximately 1.5 liters per minute of flow. Typically, three major hepatic veins are present: the right, the middle, and the left. These veins form the internal divisions between the interlobar and intersectoral planes (Plate 69). The right hepatic vein delineates the anterior from the posterior segments of the right lobe; the middle hepatic vein runs along the main fissure dividing the right and left lobes; the left hepatic vein divides the left lateral segments from segment IV. Other minor or short hepatic veins drain directly into the retrohepatic inferior vena cava from the caudate lobe and posterior segments of the right lobe.

BILIARY ANATOMY

Biliary drainage of the liver can be divided into extrahepatic and intrahepatic components. The extrahepatic bile ducts include the confluence of the right and left hepatic ducts at the hilum of the liver. The main bile duct is typically approximately 6 to 8 mm in diameter. The upper segment, called the common hepatic duct, is situated above the junction of the cystic duct, and below that point forms the common bile duct. Intrahepatically, the liver is drained by the right and left hepatic ducts. These course into the liver along with the portal vein and hepatic artery, invaginating Glissen's capsule at the hilum. Bile ducts usually lie anterior to the corresponding portal branches.

The biliary and vascular anatomy of the left lobe of the liver is unique. The left portal branch typically courses along the base of the left medial segment or hilar plate to the umbilical fissure. At that point, it divides into segments II and III, with a retrograde branch to the left medial segment (segment IV; see Plate 69). The right hepatic duct divides into posterior and anterior branches, and then subdivides superiorly and inferiorly to the four segments of the right lobe. The caudate lobe (segment I) has its own biliary and portal drainage, typically comprising two portions, the left and right processes of the caudate lobe.

RESECTION OF THE LIVER

The nomenclature for hepatic surgery can be confusing and varied in the literature. Generally, hepatic resections can be divided into two types: those defined by functional anatomical structures, and the atypical or nonanatomical resections that are not based on liver segmental structures. The most common anatomical resection is the hepatic lobectomy, either left hepatic lobectomy (which includes segments II, III, and IV) or right lobectomy (which is the resection of segments V, VI, VII, and VIII). The extended right lobectomy or right trisegmentectomy includes the left medial segment (IV) with the right lobe. The extended left lobectomy or left trisegmentectomy includes segments V and VIII with the left lobe. Either right or left extended lobectomy may or may not include all or part of the caudate lobe. More limited anatomical resections of one or two segments can also be performed, including the left lateral segmentectomy (segments II and III).

Plate 70.
Liver—Sectional Views

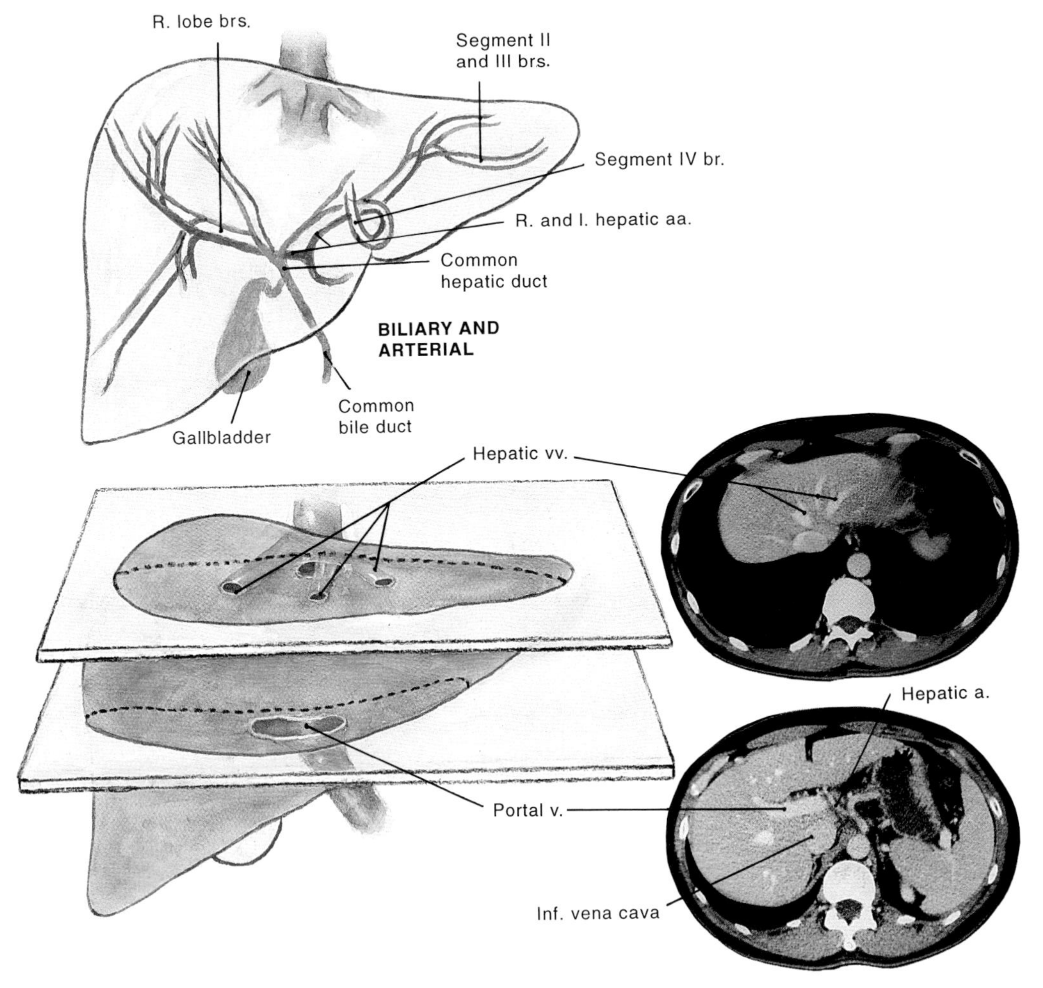

RADIOLOGIC ANATOMY

Visualization of the liver using imaging modalities including ultrasound, cholangiography, and computerized tomography can clearly demonstrate important vascular and biliary structures within the liver. Computerized tomography is able to demonstrate vascular structures within the liver as well as organs surrounding the liver. Intravenous contrast is used to aid in the visualization of vascular structures, and the use of high-resolution helical CT scan employing rapid intravenous injection and thin contiguous slices has greatly improved the detail of the intrahepatic cross-sectional anatomy (see Plate 70).

The Female Generative Tract and Pregnancy

Howard W. Jones, Jr., M.D.

The female generative tract consists of the vulva, vagina, uterus, two Fallopian tubes, and two ovaries, together with their ligamentous and muscular support and vascular and nerve supply.

The greatest part of the vulva consists of two vertical outer folds of skin (the labia majora), which converge and join anteriorly with the mons veneris, a pad of tissue overlying the symphysis pubis. Two smaller vertical folds (the labia minora), located within the labia majora, surround the vaginal orifice. Anteriorly, the labia minora are continuous with the foreskin (prepuce) of the clitoris, a small organ composed of erectile tissue and homologous with the male penis. The urethral meatus is between the clitoris and the vaginal orifice.

Entering the vulva are the ducts of Skene's and Bartholin's glands, which furnish lubrication. The hymen is a rudimentary structure at the vaginal orifice.

The vagina is a potential space about 10 cm in length extending from the vulva to the uterus. The walls of the vagina consist of an inner epithelial layer, a muscular layer, and an outer elastic fibrous layer. The vagina receives semen from the male during intercourse and provides an egress for the menstrual flow and a canal for the passage of the fetus during childbirth.

The uterus is a pear-shaped organ about 10 cm in length and 6 cm in width. Its lower portion, or cervix, extends into the upper vagina for a distance of about 2 cm. The cavity of the uterus, a triangular area about 2.5 cm on a side, is continuous with the cervical canal, and by this with the vagina below and with the lumina of the Fallopian tubes above. The uterus, whose muscle wall is about 2 cm in thickness, is held in position by the broad ligaments. The lining of the uterus (endometrium) undergoes changes each month in preparation for pregnancy. In the event pregnancy does not occur, the endometrium is discharged as the menstrual flow. The fertilized ovum enters the endometrial cavity from seven to ten days after ovulation and there implants itself within the endometrium, where it grows and develops.

The Fallopian tubes, each about 10 cm in length, are held in position by the broad ligaments. Each tube consists of a narrow isthmic portion attached to the uterus, a longer ampullary portion, and a distal infundibular portion terminating in the fimbria, which collect the egg after ovulation.

The ovaries are two bodies about 3.0 by 2.5 by 2.0 cm that weigh about 10 gm. They are attached to the infundibulopelvic ligaments, which contain the ovarian vessels. At birth, each ovary contains about 200,000 developing oöcytes, of which about 200 are ovulated during menstrual life.

In addition to acting as the reservoir of genetic material for future generations, the ovary is an important endocrine gland, which produces the female sex hormone (estrogen), as well as some androgen, and progesterone from the corpus luteum, which develops from the ovarian follicle after ovulation and has a life span of about fourteen days.

INNERVATION OF THE UTERUS, CERVIX, VAGINA, BLADDER, RECTUM, AND PERINEUM

The pelvic viscera are innervated mostly by the sympathetic nervous system. However, somatic sensory and motor pathways supply the urinary and intestinal voluntary sphincters, as well as the skin and muscles of the perineum.

The anatomic arrangement of the sympathetic nerves varies widely. Retroperitoneally, a network of sympathetic fibers in front of the aorta is known as the aortic plexus. On either side of this are two main strands of nervous tissue derived from the lumbar sympathetic ganglia. These fuse with the aortic plexus in front of the fifth lumbar vertebra or thereabouts to form the superior hypogastric plexus from which courses the hypogastric (presacral) nerve.

In front of the upper part of the sacrum, the hypogastric nerve becomes the middle hypogastric plexus; it then divides into two parts around the rectum, where it becomes known as the inferior hypogastric plexus. This plexus passes forward onto the uterosacral ligament and is known as the ganglion or plexus of *Frankenhäuser,* the *lateral cervical plexus,* the *uterovaginal plexus,* or, more commonly, the *pelvic plexus.*

The pelvic plexus is made up of interlacing nerve fibers containing large numbers of small microscopic ganglia. The pelvic plexus receives fibers from the sacral sympathetic ganglia and parasympathetic fibers from the nervi erigentes. The sympathetic fibers relay in the ganglia, but the parasympathetic fibers merely pass through the cell stations to the adjacent viscera.

The parasympathetic nerves supplying the pelvic viscera (except for the ovaries and the proximal part of the tubes) arise from the anterior roots of the second, third, and fourth sacral nerves. These nerves fuse and form the nervi erigentes.

The external genitalia have a parasympathetic and sympathetic nerve supply from the same sources as the pelvic organs. The parasympathetic supply dilates the vessels to the erectile tissue and causes the bulbocavernosus and the ischiocavernosus muscles to contract with sexual orgasm.

Somatic sensory branches to the skin of the perineum, including parts of the external genitalia and motor branches to the muscles of the perineum, are derived from the pudendal and ilioinguinal nerves.

The innervation of the uterus is a subject of some controversy. Apparently the uterus has a sympathetic supply from the hypogastric nerve and a parasympathetic supply from the sacral plexus. The sympathetic supply seems to be responsible for controlling the circular muscular fibers around the cervix.

Sensory fibers pass to the spinal cord from the uterus through the presacral nerve. Presacral sympathectomy severs these pain-conducting fibers, useful for the relief of spasmodic dysmenorrhea and resulting in a painless first stage of labor.

The Fallopian tubes are innervated peripherally by sympathetic fibers from the ovarian plexus and medially by sympathetic fibers derived from the pelvic plexus. The parasympathetic supply is derived from the vagus via the ovarian plexus.

Both the sympathetic and parasympathetic fibers to the ovary reach the ovary by way of the ovarian plexus.

The bladder has a sympathetic supply from the hypogastric nerves and a parasympathetic supply through the nervi erigentes.

The anatomy of the pelvic floor is shown in Chapter 27, in association with the perineum and male genitourinary system.

PREGNANCY

After fertilization, which takes place in the Fallopian tube, the early embryo enters the uterine cavity and buries itself into the endometrium, where it first develops. All organs of the embryo develop from the three primary germ layers (ectoderm, mesoderm, and endoderm), which form early in the embryo. From the ectoderm are derived the nervous system and sense organs, the epidermis, brain, and spinal cord. From the mesoderm are derived the skeletal, muscular, circulatory, and some parts of the reproductive system and the kidneys, ureters, and other structures; and from the endoderm are derived the alimentary canal and its derivatives, such as the respiratory system, thyroid, pancreas, liver, gallbladder, and parts of the reproductive organs.

During pregnancy, the uterus gradually rises in the abdomen, until near the end of pregnancy it reaches almost to the xiphoid. Before labor, the head drops lower into the pelvis, and the fundus of the uterus, prior to labor, may be somewhat lower than previously.

A great variety of fetal abnormalities occur, some of which are genetic, others environmental, but much remains to be learned about this aspect of human development. However, certain types of genetic defects can be ascertained during embryonic life. This can be done by the examination of cells removed with a small amount of amniotic fluid at about the sixteenth week of gestation. The cells in the amniotic sac are of fetal origin, and after they have been cultured they may be examined for chromosomal content and certain biochemical constituents, abnormalities of which reflect the diseased state of the fetus. At the present time, the only method of control is termination of the pregnancy of mothers carrying seriously handicapped children, e.g., those affected with Down's syndrome (47, XX, +G21 or 47, XY, +G21) or Tay-Sachs disease, a very serious disorder due to an enzymatic deficiency inherited as an autosomal recessive.

At term, labor is initiated by a mechanism as yet unknown. During the first stage of labor, uterine contractions press the presenting part, usually the head, into the cervix as the cervical canal dilates.

During the second stage of labor, uterine contractions become more frequent and painful, and at this stage the abdominal muscles by voluntary action may aid in propelling the fetus through the vagina. Immediately after birth, the umbilical cord can be ligated and severed.

In the third stage of labor, the placenta is expelled. With the delivery of the child and the placenta, the uterus contracts and thereby stops the bleeding that would otherwise occur. Contractions of the uterus may be slightly painful for several days and are known as *afterpains.*

Plate 71.
Genitourinary Tract, Female

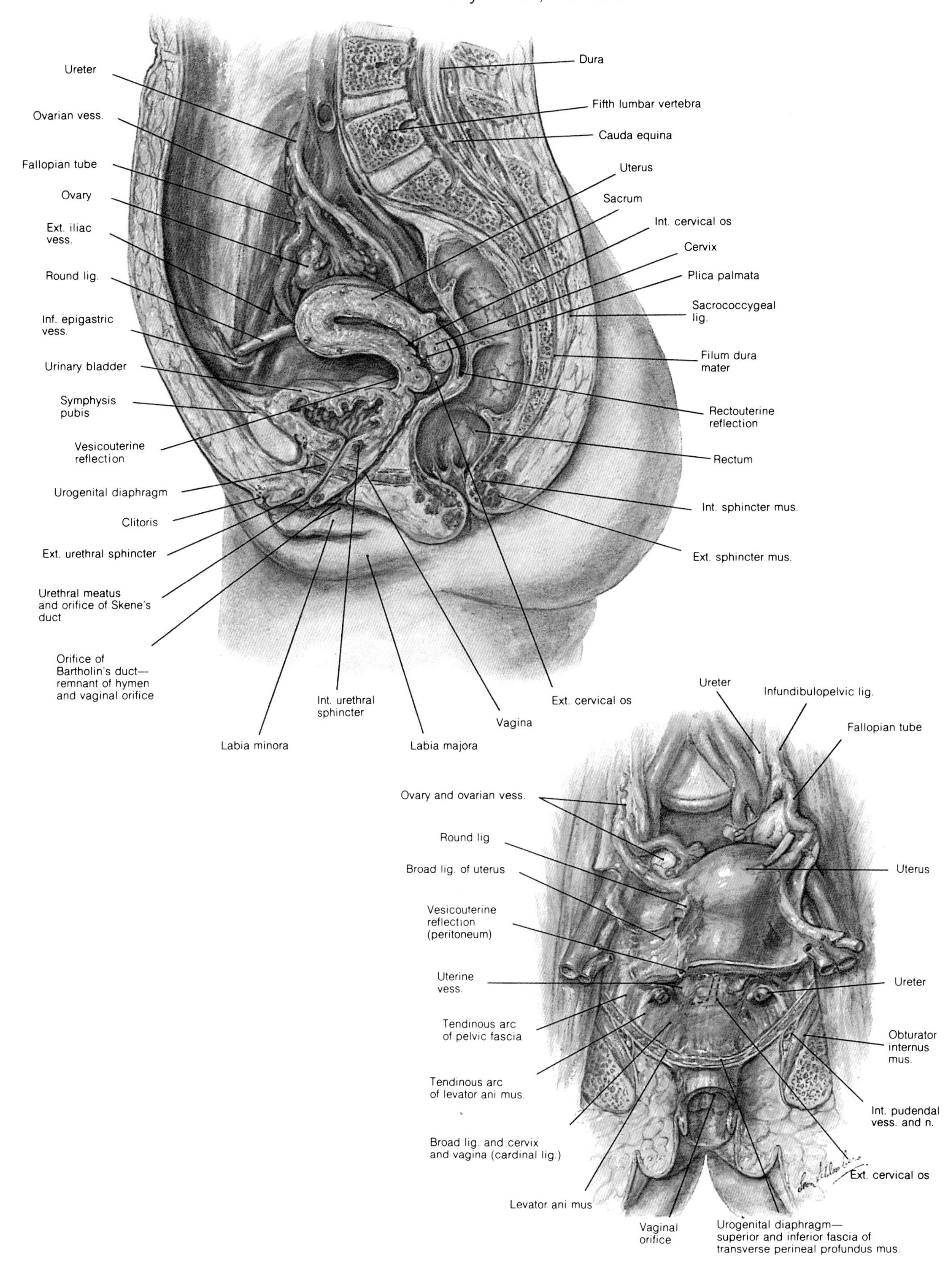

Plate 72.
Genitourinary Innervation of the Uterus, Cervix, Vagina, Bladder, Rectum, and Perineum

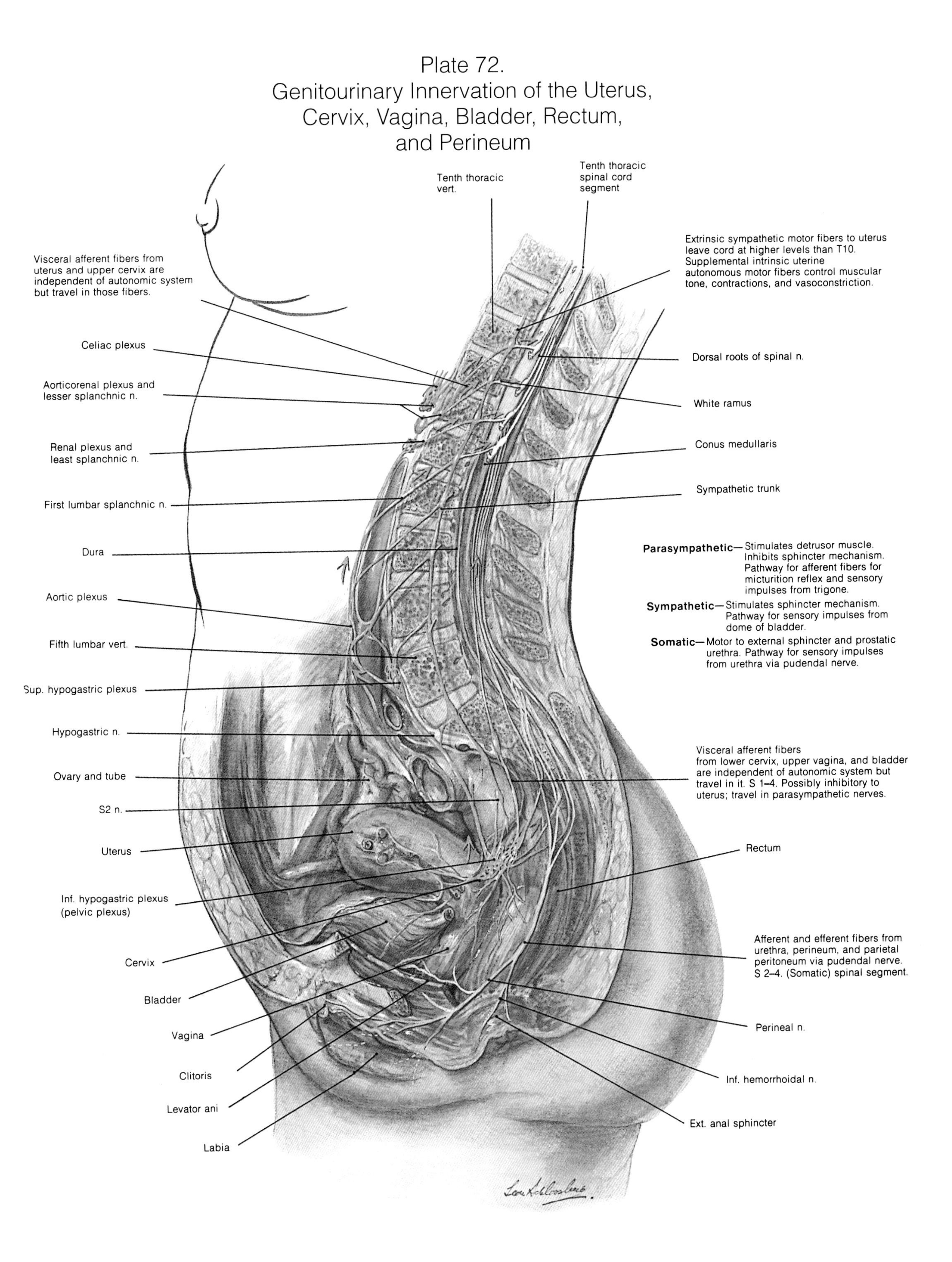

Plate 73. Pregnancy

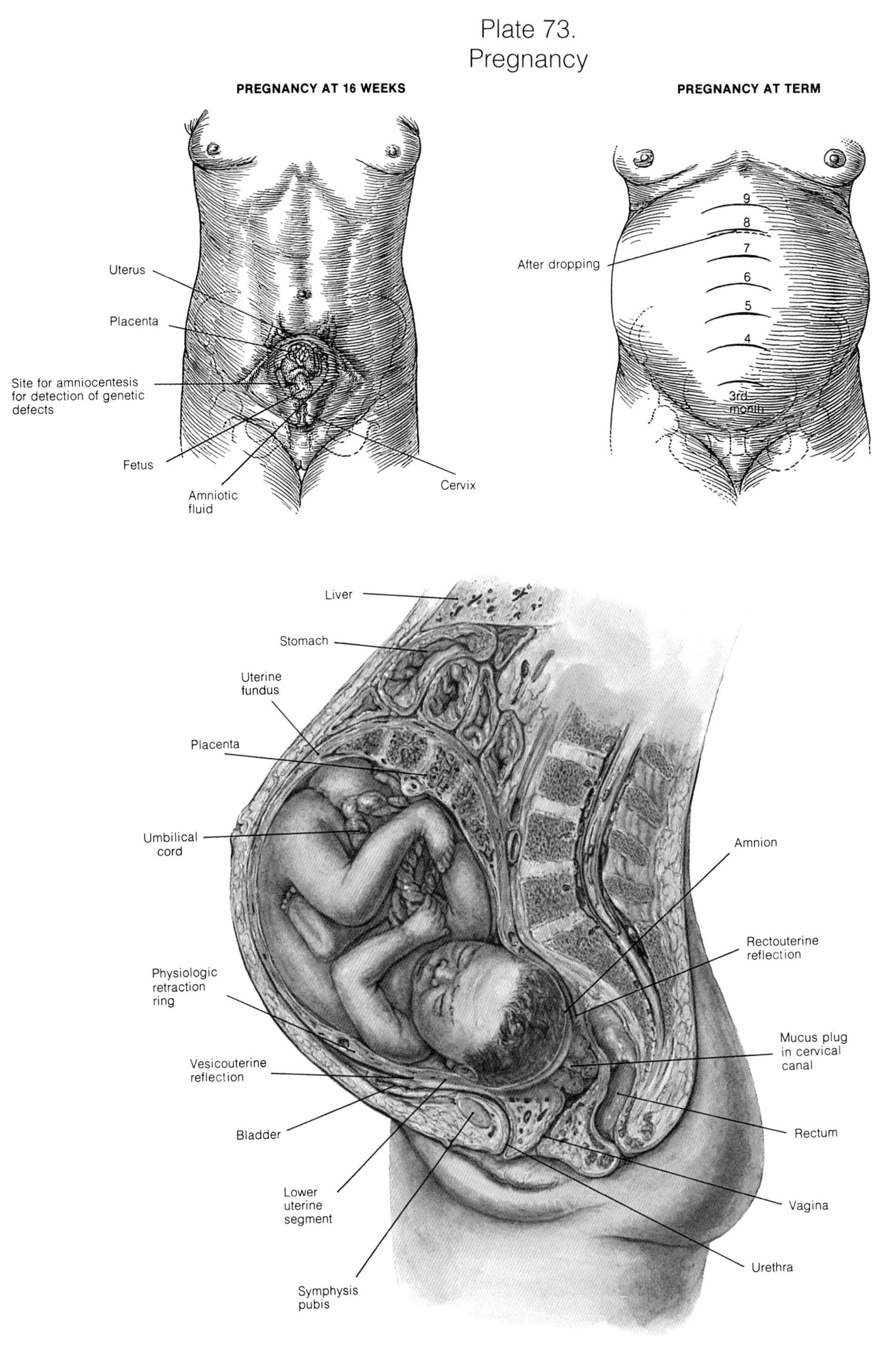

PREGNANCY AT TERM

The Menstrual Cycle

H. Lorrin Lau, M.D.

The term *menstruation* refers to the monthly shedding of endometrium from the uterus, whereas the term *human menstrual cycle* refers to a recurring series of events in the hypothalamic-pituitary-ovarian-uterine axis responsible for ovulation and reproduction.

Day 1 of the cycle is the first day of bleeding.

Changes in geophysical location; influences such as light, smell, and sound; and emotional stress such as death in the family or the stress of college examinations can seriously disturb the rhythm of the cycle or cause a period of amenorrhea. These external stimuli act presumably via the cortex to the hypothalamus.

The hypothalamus produces releasing factors (peptides of MW 2000) for follicle-stimulating hormone (FSH), luteinizing hormone (LH), and a prolactin-inhibiting factor (PF). These are carried by the hypophyseal portal capillary network to the anterior pituitary, where the release of FSH and LH is stimulated but that of prolactin is inhibited. This initiates the follicular phase, in which there is growth of the Graafian follicle, proliferation of endometrium, rises in serum FSH and LH, and increased secretion by the theca interna of 17 α-hydroxyprogesterone, androstenedione, and estradiol. Steroidal contraceptives such as norethynodrel with mestranol (Enovid) act on the hypothalamus by depressing the releasing factors.

At midcycle, or the ovulatory phase, which is day 14 in the idealized cycle, LH induces final maturation of the follicle ripened by FSH and expulsion of the egg from the surface of the ovary. FSH and LH show their largest spike at this time (day 14); a smaller spike of FSH occurs in the follicular phase in many cycles. The space left by the egg becomes a corpus luteum (yellow body), which secretes progesterone, the steroid that resets the hypothalamic thermostat to give the typical elevation in the pattern of basal body temperature. Other steroids such as estradiol and 17 α-hydroxyprogesterone rise concomitantly with the LH peak. By contrast, FSH alone does not stimulate the production of steroids.

The life span of the corpus luteum is about fourteen days, determines the duration of the secretory phase, and is affected by luteotrophic and luteolytic agents. LH is definitely luteotrophic in man, while prolactin may also be luteotrophic. In sheep, prostaglandins (lipid-soluble, unsaturated hydroxy acids with 20 carbons) are definitely luteolytic, while in man their luteolytic activity is still being studied.

The uterus reflects the steroidal activity of the ovary. In the proliferative phase, marked mitotic activity is seen in the glands and stroma with straight narrow glands containing basal nuclei and pseudostratification. At ovulation, these decrease. After ovulation basal vacuoles appear in the apex of the glands, which enlarge and acquire a sawtooth appearance, with ragged edges, where secretion is discharged into the lumen. The stroma swells and spiral arteries become prominent. Two days prior to menstruation, the corpus luteum regresses to induce disappearance of the stroma swelling, spasm of the arteries, and a filling with material of the glands, which show an irregular apex and single round basal nuclei without stratification.

The vaginal epithelium, thickened at ovulation with cornified cells, decreases in height and shows precornified cells.

Spasm of the spiral arteries, hypoxia of the endomentrium, and the appearance of sheets of predecidual cells initiate sloughing of the uterine epithelium (*menses*), and another cycle begins.

Plate 74.
The Menstrual Cycle

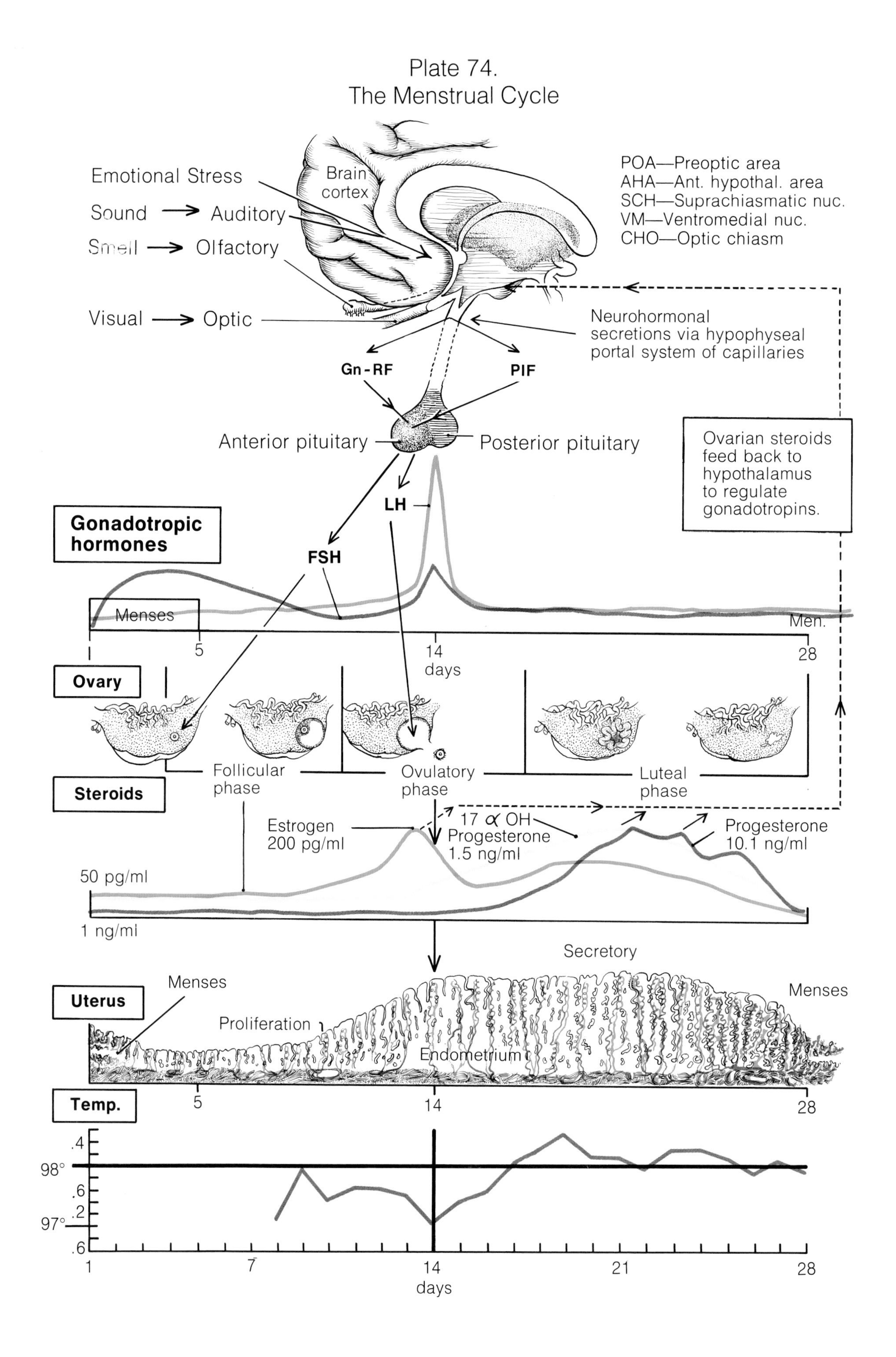

27 The Kidneys, Male Genitourinary System, and Perineum

Rainer M. E. Engel, M.D.

The normal urinary tract consists of two kidneys, two ureters, a bladder, and a urethra. Except for the urethra, the urinary tract is essentially the same in both males and females. The function of the urinary tract consists of maintenance of fluid and electrolyte balance, which is achieved by secreting water and various waste products of the body. A number of substances are conserved by reabsorption in the kidney. Others are excreted, and the final end product, urine, is delivered into the collecting system.

The kidneys are paired organs, each approximately 11 cm long and 6 cm wide, that lie in the retroperitoneal area at the level of the lower thoracic and upper lumbar vertebrae. The right is usually somewhat lower. The upper pole borders on the diaphragm, and the lower portion extends over the iliopsoas muscle. The posterior surface is protected in its upper part by the lower ribs. The renal tissue is covered by the renal capsule and surrounded by Gerota's fascia, which is quite firm and will usually confine blood and urine extravasations as well as suppurative processes. Medially, blood vessels, lymphatics, and nerves enter each kidney at its midportion, the hilum. Behind the blood vessels, the renal pelvis, with the ureter, leaves the kidney. Blood enters the kidney through the renal artery, which is usually single, and which branches into smaller vessels supplying the different lobes of the kidney. The kidneys receive about one-fourth of the cardiac output per minute. Once the artery has entered the renal substance, it branches along the boundary between cortex and medulla and from there radiates into the parenchyma. There are no communications between the capillaries or larger vessels of the kidney. The arcuate arteries supply the cortex and give off small arterioles that form multiple convoluted tufts, the glomeruli. From each glomerulus the efferent, still arteriolar vessel leaves again to supply, with a fine network, the renal tubule of its corresponding glomerulus. These peritubular arteries empty through small venules into larger collecting veins and finally, through the renal vein, into the vena cava. The left renal vein is longer than the right, as it crosses the aorta to reach the vena cava, and receives the left gonadal vein. The right gonadal (ovarian or spermatic) vein empties separately, below the renal vein, into the vena cava.

Numerous lymphatic channels are found throughout the kidney. These drain into hilar nodes, which communicate with periaortic nodes above and below the hilar area. Cross communications to the contralateral side have also been demonstrated.

Urine is filtered by the glomerulus and collected into a space confined by Bowman's capsule. From there it is transported through the proximal convoluted tubule, Henle's loop, and distal convoluted tubule into the collecting tubules that empty through the pyramid of the medulla into the caliceal cups. Urine is filtered mainly by the hydrostatic pressure of the blood pressure. Thus, when the blood pressure drops, filtration will also stop and urine formation ceases. Other factors important in the formation of urine are: (1) the osmotic pressure, which is largely produced by the plasma protein in the blood; and (2) the back pressure of already excreted urine in the collecting system. The glomerulus acts, in other words, like a sieve that will strain corpuscles and will also hold back protein. This glomerular filtration would allow approximately 190 liters of fluid to be excreted daily. However, as the filtrate passes from the glomerulus into Bowman's capsule and into the tubules, reabsorption, secretion, and excretion will alter the final end product. Only about 1 percent of the total filtrate will be excreted as urine into the renal pelvis.

Hormones play an active role in the reabsorption of both water and other substances. Antidiuretic hormone (ADH) regulates absorption and elimination of water, depending on the needs of the body. Aldosterone promotes reabsorption of sodium and excretion of potassium. Parathyroid hormone increases the reabsorption of calcium and decreases the reabsorption of phosphorus.

The amount of functioning renal tissue is fortunately far in excess of the minimum requirements for life. About one-third of the normal tissue will adequately, and without appreciable alteration of function tests, sustain life and growth.

After urine has entered the collecting system, it remains unchanged. The urine is collected in the renal pelvis and moves, by peristaltic waves, across the ureteropelvic junction and through the ureter. One of the common sites of obstruction of kidneys is at the level of the ureteropelvic junction. The blood supply to the ureter is derived from numerous areas. Fine branches having their origin from the renal blood vessels supply the ureter from the renal pelvis. The lower portion receives its blood supply from vesical arteries, and the midportion is supplied by branches from the lumbar vessels. The lymphatics drain into the areas that correspond to the arterial supply and veins show a similar distribution. The ureters enter the bladder through a long tunnel through the muscular wall of the bladder and the mucosa. Each ureteral orifice is a small slitlike opening. The ureters usually lie about 2 to 3 cm apart in the adult and are situated slightly off the midline, about 2 cm above the internal opening of the urethra. The area between these three openings is called the *trigone*. Under normal conditions, urine will pass though the ureteral orifice only in one direction, i.e., into the bladder. As the bladder pressure increases, the mucosal tissue over the inner wall of the ureter will be pressed against the back wall of the ureter, thus preventing backing up of urine, or vesicoureteral reflux. As the ureter passes from the kidney into the bladder, it encounters three narrow points. The first is at the ureteropelvic junction; the second, at its crossing with the iliac vessels; and the third, where it penetrates the bladder wall. Stones, during their passage from the kidney down into the bladder, may lodge at one of these three points and produce obstruction.

The bladder is a rounded, hollow muscular organ that normally distends to hold an average of 500 ml. However, under certain conditions, the bladder can be distended much beyond this capacity. In the male, the posterior surface of the bladder is in proximity to the rectum. In the female, the superior part of the vagina and the uterus are interposed between the bladder and rectum. The dome of the bladder is covered by peritoneum.

The bladder receives its blood supply from branches of the internal iliac or hypogastric arteries, with smaller branches from the hemorrhoidal and uterine arteries. Lymphatic drainage, which is important in the spread of bladder cancers, follows the internal, external, and common iliac vessels predominantly.

The nerve supply to the bladder includes the parasympathetic system, which supplies the detrusor muscle, which will contract the bladder; the sympathetic branch of the autonomic system supplies the base of the bladder. The pudendal nerve supplies the external sphincter, which surrounds the urethra. Connections, or synapses, between these various nerve supplies allow for simultaneous contraction of the detrusor, and relaxation and opening of internal and external sphincters. Sensory fibers that transmit both filling and stretch sensation of the distended bladder are carried through parasympathetic fibers to the spinal cord, where the primary reflex center for the bladder is located at the level of S2 to S4. A reflex arc may, at this level, permit some form of function of the bladder in certain patients with spinal cord injuries. Tracts in the spinal cord connect the primary center with higher centers that allow us to suppress the urge to void and become "toilet-trained." Thus, the normal bladder will continue to fill without causing us discomfort, and at its usual filling limit will elicit nervous stimuli that we, however, can override, to expand the capacity and empty the bladder at our convenience.

The ureters will permit transport of urine into the bladder. Even with complete filling of the bladder, there will be no incontinence of urine. Once the act of voiding or micturition begins, the bladder will empty to completion.

Urine leaves the bladder through the urethra. In the female, this is a fairly short tubular organ about 3 to 5 cm long with its external opening between the labia minora; it courses along the anterior vault of the vagina. The male urethra is an S-shaped, tubular organ, approximately 20 cm long. At its beginning, it runs through the prostate, a secondary sex gland. The prostatic urethra is 2.5 to 3.0 cm long. Just below the prostate, the urethra pierces the pelvic diaphragm, an area in which it is almost immobile and not very distensible. This diaphragmatic portion of the urethra is also called the *membranous urethra,* and is approximately 1 cm long. Below this, the bulbus urethra, a patulous part of the urethra, begins, and, following this, the channel tapers at the penoscrotal junction into the pendulous urethra, which lies in the ventral wall of the penis and is on its ventral surface, covered by the corpus spongiosum.

The bladder neck is the most common site of obstruction of the urinary tract in the male. Usually this is produced by prostatic enlargement, due to benign or malignant processes. As the prostate enlarges, it not only grows toward the outside perimeter but also compresses the lumen of the urethra. In benign prostatic enlargement, the small periurethral glands enlarge to form an adenoma. This can be removed by different types of prostatectomies; the true prostatic tissue in these operations is left intact. The true prostatic glands also empty into the prostatic urethra, via a dozen small

Plate 75.
Genitourinary Tract, Male

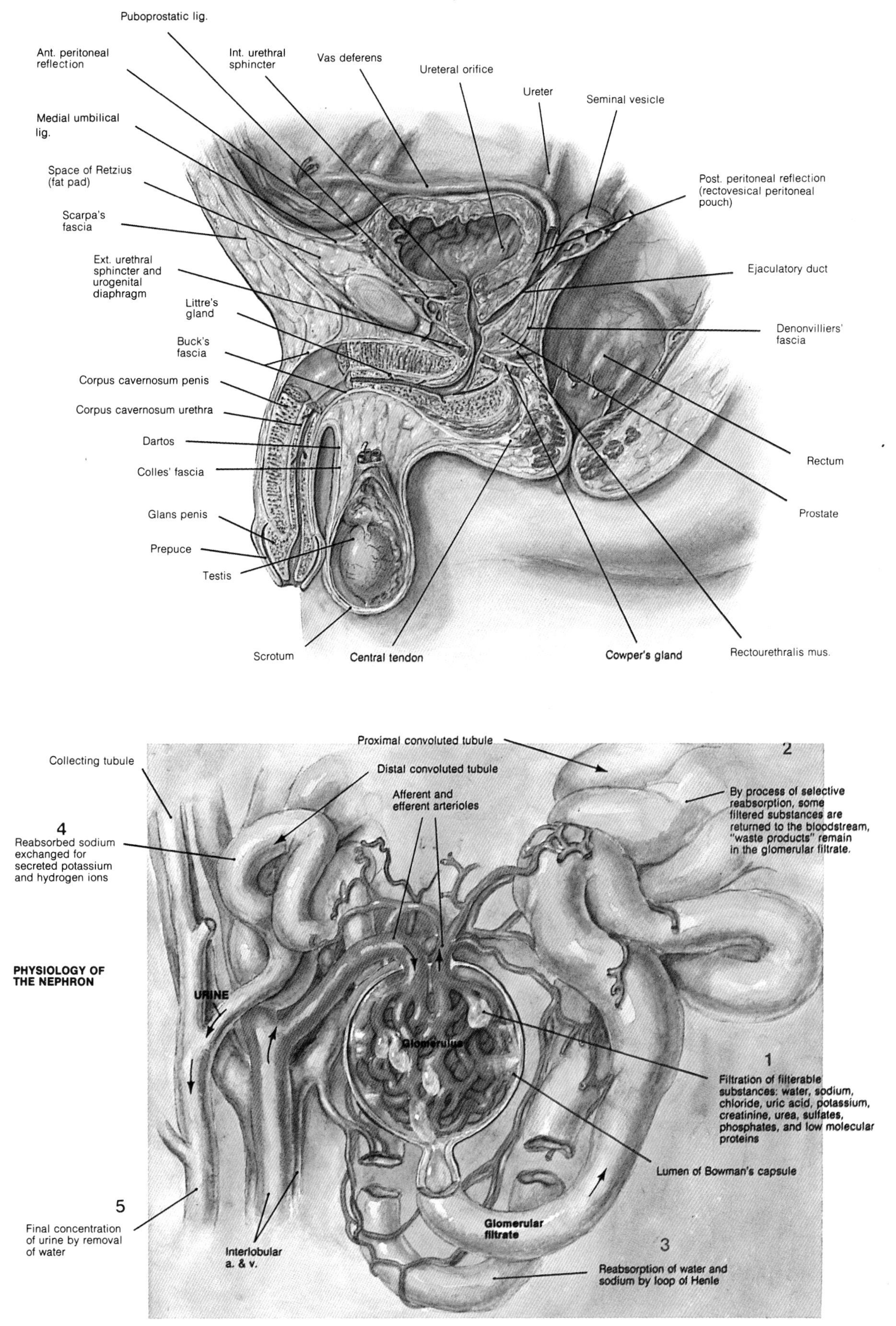

Plate 76.
Genitourinary Tract, Male—Vessels and Nerves

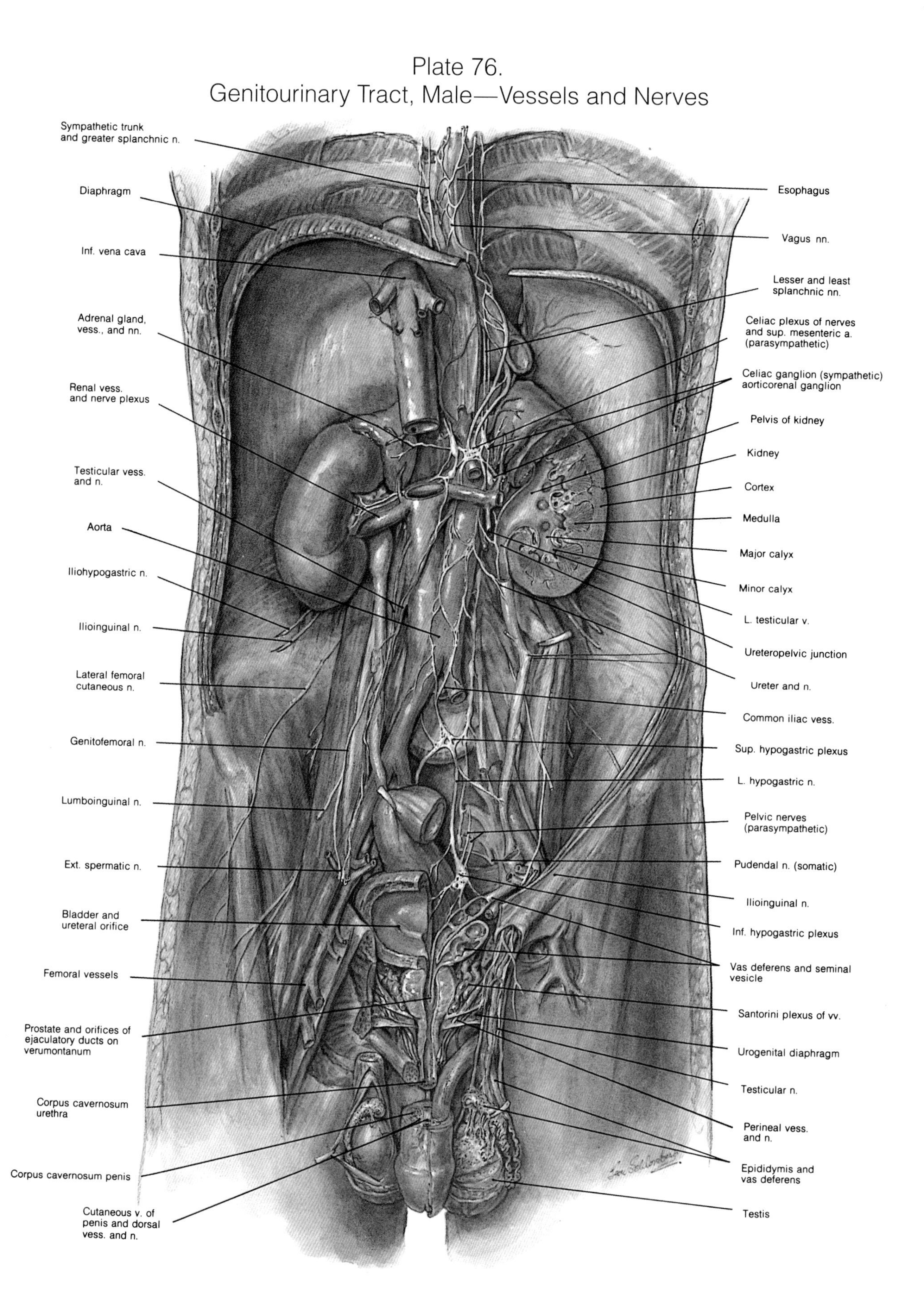

ducts, into the area of the verumontanum. The two ejaculatory ducts also open into this area. The paired Cowper's glands secrete a small amount of fluid, which enters the urethra at the pelvic diaphragm. Scattered along the remainder of the urethra are the numerous small glands of Littre. Occasionally they harbor infection.

The male genital tract comprises the testes and epididymides, which lie in the scrotum and lead into the vas deferens. The vas deferens is a tube-like structure that passes through the inguinal ring, lateral to and then behind the bladder, where, after forming the ampulla, it joins with a small duct from the seminal vesicle into the ejaculatory duct.

The ejaculatory duct traverses the prostate and opens into the prostatic urethra. During delivery of the ejaculate, the combined secretion of the testes, seminal vesicles, and prostate is propelled through the urethra. At the time of ejaculation, the bladder neck stays closed, the external sphincter opens, and thus the ejaculate is propelled outwards. In patients who have undergone a prostatectomy or resection of the bladder neck, the area of least resistance is toward the bladder, and they therefore may experience a dry ejaculation or retrograde ejaculation into the bladder.

The blood supply to the testis comes from the testicular artery, which originates on the left side from the renal artery and on the right side directly from the aorta just below the renal artery. The high origin of these vessels is explained by the embryologic origin of the testes in this area. Incomplete descent can lead to intra-abdominal retention of the testis. Venous drainage occurs along the spermatic veins, which parallel the arteries.

The function of the testes is twofold: (1) they produce the male hormone, testosterone; and (2) they produce spermatozoa, which travel from the tubules of the testes into the epididymis, where they undergo maturation. From there, they are delivered into the vas deferens. Thus, a vasectomy will only interrupt delivery of spermatozoa. The bulk of the ejaculate is composed of the fluid of the secondary sex glands, i.e., the seminal vesicle and prostate. This is not affected by vasectomy.

The urethra thus serves a twofold purposes, i.e., as a passageway for both urine and the ejaculate.

Erection of the penis is achieved through filling of the three expansile bodies of the penis with blood. These include the corpus spongiosum, which is on the undersurface of the urethra, and the paired large corpora cavernosa, which are anchored at the pubic rami and receive their blood supply from the pudendal arteries. Under erogenous stimulation, the outflow of these bodies is partially closed, and the resultant infusion of blood produces the necessary rigidity. This stimulation is mediated through branches of the sympathetic and parasympathetic nervous system, although most of the stimulation is of cerebral origin. The seminal vesicles and the prostate gland deliver fluid that contains nutrients and improves the motility of the spermatozoa. Of these, the most important is the prostate. Its blood supply is derived from branches of the inferior vesical artery, and it drains into a rich plexus of veins, the most important of which is the plexus of Santorini, on the anterior surface of the prostate.

The anatomy of the male and female perineum and pelvic floor is shown in Plate 77. Of interest is the course of the internal pudendal nerve which exits from Alcock's canal where it can be anesthetized in an effort to decrease urinary outflow resistance in patients with neurogenic imbalance between the detrusor and the musculature of the external urethral sphincter. More permanent interruption of function can be achieved by removing a portion of the middle branch, leaving intact the nerve supply to the anal sphincter and penis.

The portion of the top drawing showing dissection of the penis serves to illustrate how it is possible to separate the spongiosum from the corpora cavernosa, an anatomical fact that is helpful in partial amputation of the penis for cancer.

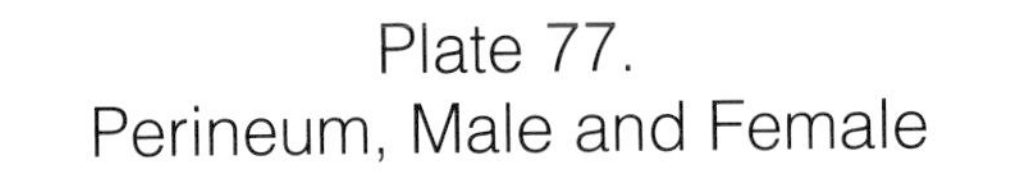
Plate 77. Perineum, Male and Female

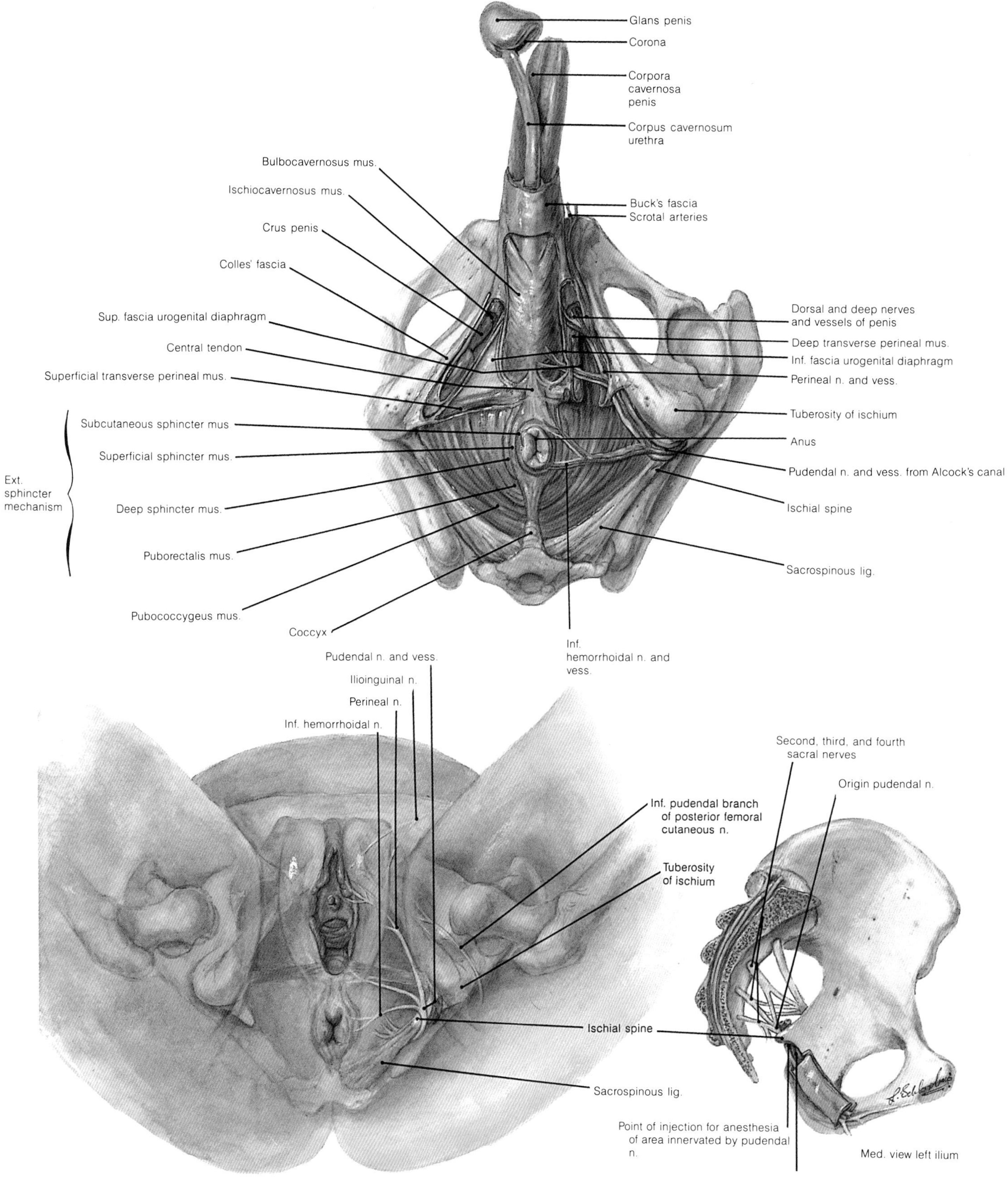

The Prostate and Male Pelvis

Patrick C. Walsh, M.D.

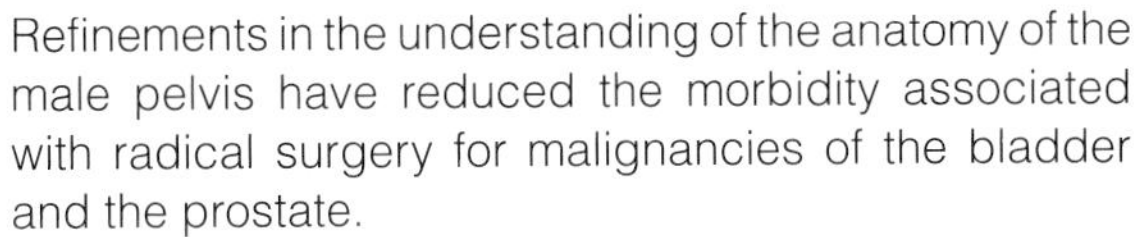

Refinements in the understanding of the anatomy of the male pelvis have reduced the morbidity associated with radical surgery for malignancies of the bladder and the prostate.

The prostate receives its arterial blood supply from the inferior vesical artery, which provides small branches to the seminal vesicles and the base of the bladder, then terminates in two large groups of prostatic vessels: the urethral and capsular groups. The capsular branches run along the pelvic sidewall in the lateral pelvic fascia posteriorly, providing branches that course ventrally to provide the outer portions of the prostate. These capsular vessels provide the macroscopic landmark that aids in the identification of the microscopic branches of the pelvic plexus that innervate the corpora cavernosa (neurovascular bundle). The deep dorsal vein leaves the penis under Buck's fascia between the corpora cavernosa and penetrates the urogenital diaphragm, where it travels over the ventral and lateral surfaces of the prostate, terminating in the internal pudendal vein.

The autonomic innervation of the pelvic organs and external genitalia arises from the pelvic plexus, which is formed by parasympathetic visceral efferent preganglionic fibers that arise from the sacral center (S2 to S4) and sympathetic fibers from the thoracolumbar center (T11 to L2). The pelvic plexus is located retroperitoneally beside the rectum, 5 to 11 cm from the anal verge, and forms a fenestrated rectangular plate that is situated in the sagittal plane with its midpoint located at the tip of the seminal vesicle. The pelvic plexus provides visceral branches that innervate the bladder, ureter, seminal vesicle, prostate, rectum, membranous urethra, and corpora cavernosa. The nerves innervating the prostate travel outside the capsule of the prostate and Denonvillier's fascia until they perforate the capsule and enter the prostate. The branches to the membranous urethra and corpora cavernosa also travel outside the prostatic capsule in the lateral pelvic fascia dorsolaterally between the prostate and rectum in the neurovascular bundle.

Although the external sphincter, at the level of the membranous urethra, is often depicted as a sandwich of muscles in the horizontal plane, the striated urethral sphincter, which is responsible for passive urinary control in men, is a vertically oriented tubular sheath that surrounds the membranous urethra. At the apex of the prostate, the fibers are horseshoe shaped and form a tubular striated sphincter surrounding the membranous urethra. The striated sphincter contains fatigue-resistant, slow-twitch fibers that are responsible for passive urinary control, whereas active continence is achieved by voluntary contraction of the levator ani musculature. The pudendal nerve provides the major nerve supply to the striated sphincter and the levator ani.

Plate 78.
The Prostate and Male Pelvis

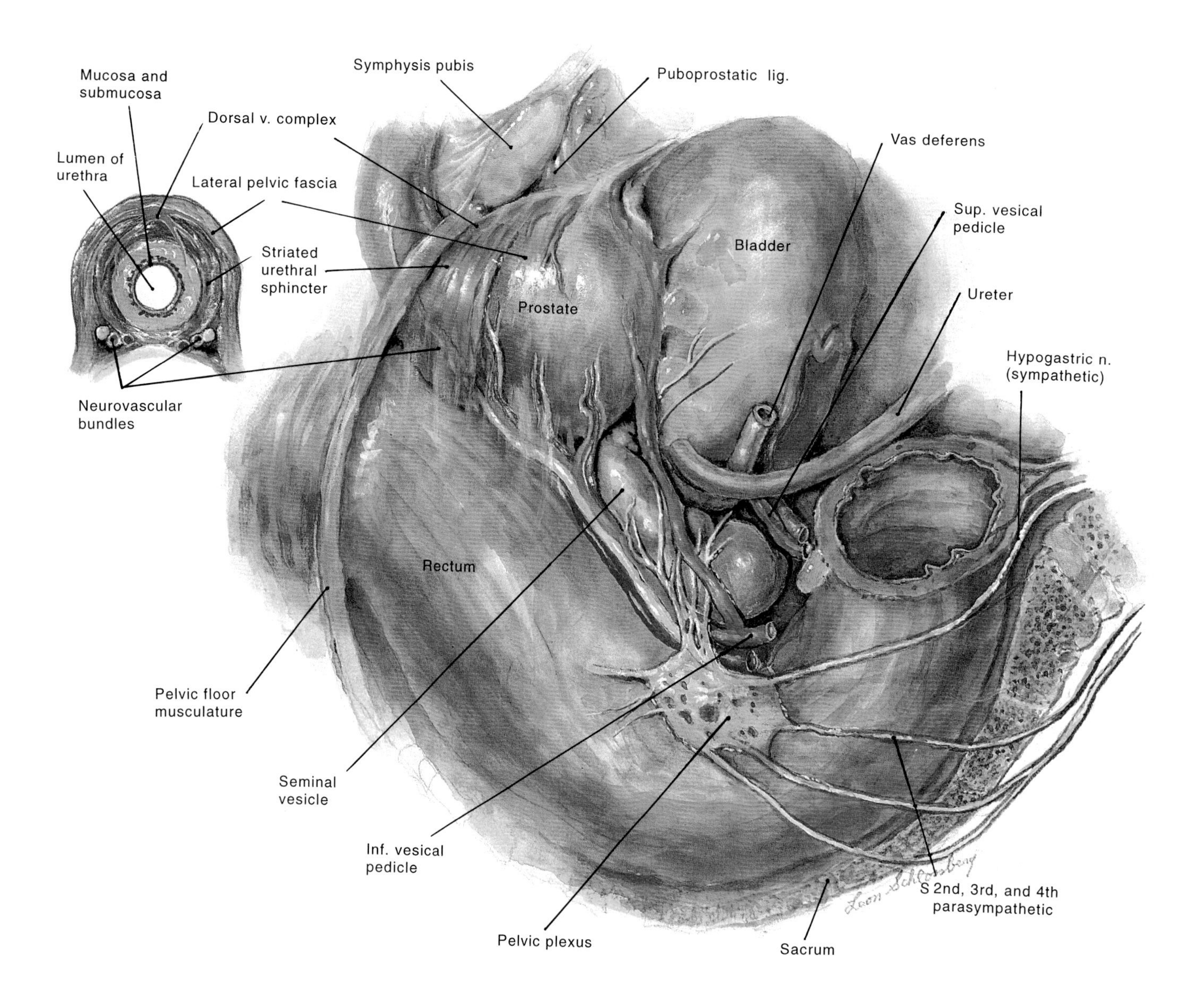

The Skin

James J. Ryan, M.D.

The largest organ of the body, the skin, is best understood in terms of its principal function as an interface between the internal and external world. The regulation of the internal environment by detection of, protection from, and adaptation to the surrounding environment is clearly reflected in the functional anatomy of the skin. There are two types of skin: hair-bearing, covering the vast majority of the body surface; and hairless, confined to the soles of the feet and the palms of the hands. The difference in the two is simply the presence or absence of the pilosebaceous apparatus, the hair follicle and the accompanying sebaceous gland. In thickness, three major tissue layers are identified. The uppermost is a thin, stratified epithelium, the epidermis. Beneath the epidermis is the dense, fibroelastic connective tissue stroma called the *dermis* or *corium*. The third layer of the skin is the subcutaneous tissue composed of areolar and fatty connective tissue.

The epidermis or cellular investment of the entire organism consists of two principal divisions: an upper, thick keratin layer of packed cells without nuclei, the stratum corneum, and an underlying layer of nuclear cells, the stratum Malpighii, histologically divided into three layers of progressive nuclear degeneration—the stratum germinativum. Two types of cells are present in the stratum germinativum, keratinocytes and melanocytes.

The former are most numerous and, as they differentiate and migrate upward, form the stratum corneum, which is renewed on an average of every twenty-six days. The melanocytes form the pigment granules, which are transferred to the keratinocyte cells and give the skin its color and much of its protection from intense light. The importance of the stratum corneum is to limit permeability of the skin to water and ions.

The dermis, a complex material predominantly composed of collagen and elastic fibers and diffuse ground substance, encloses cellular systems of nerves, vessels, glands, and appendages. There is a vast three-dimensional network of blood vessels, predominantly concerned with thermal regulation, seen as the subdermal vascular plexus, a dermal vascular plexus, and a subpapillary vascular plexus. The glands are the sweat glands, which provide a powerful physiologic mechanism for heat loss. The 2 to 3 million exocrine sweat glands distributed over the entire body surface are capable of delivering 2 to 3 kg of watery sweat per hour. A second type of sweat gland, the apocrine gland, occurs principally in the axilla, and is most responsive to nervous stimulation. The third gland is the sebaceous gland seen emptying into the shaft of the hair follicle and thus forming a part of the combined pilosebaceous apparatus. The skin surface lipids are largely derived from the sebum, serving to lubricate the skin and possibly influence its bacterial flora.

The subcutaneous tissue is principally adipose tissue, varying greatly in thickness over different areas of the body. This insulates underlying tissues from extremes of environmental heat or cold, as well as providing a substantial cushion. Strands of collagen extending from the dermis through the subcutaneous tissue to underlying muscle, fascial, or bony periosteal attachment influence the mobility of the skin. In the subcutaneous and dermal areas innumerable free and encapsulated nerve endings provide sensory input of many forms from the body surface.

Plate 79.
The Skin

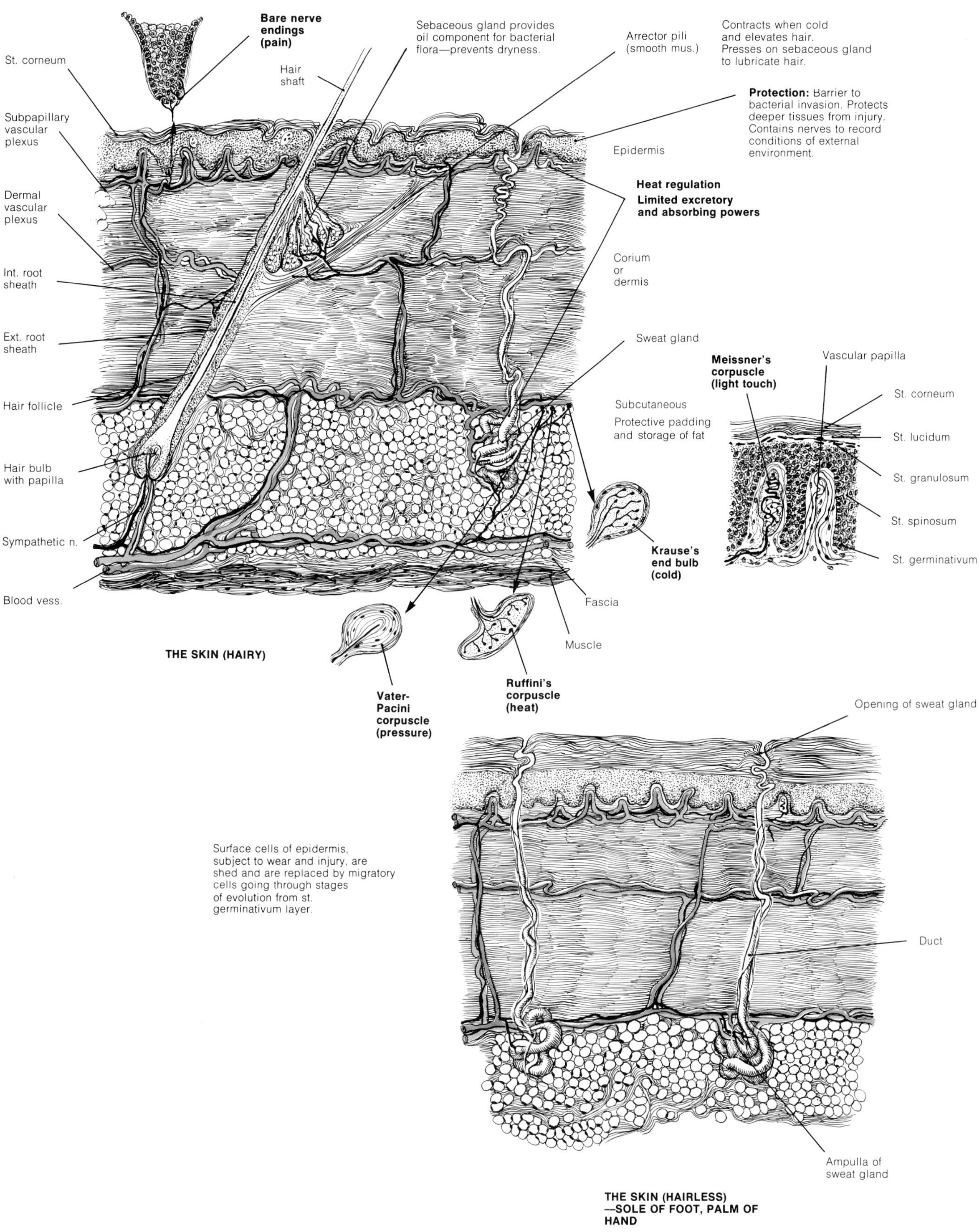

Index

Pages listed in **boldface** type contain illustrations.